six old SILENCER catalogs

compiled and edited
by
Donald B. McLean

the COMBAT BOOKSHELF

Normount Technical Publications

First printing, October, 1971
Second Printing, December, 1974

ISBN: 0-87947-006-2

Printed in the U.S.A.

Also available
from the publisher;

FIREARM SILENCERS, Vol. 1, U.S. $4
SILENCERS, Patterns and Principles $5.95

free catalog of books on modern and antique weapons is available from the publisher :

Normount Technical Publications
P.O. Drawer N-2-B
Wickenburg, AZ 85358 USA

CONTENTS:

SHOOTING WITHOUT NOISE[1]

WITH A

MAXIM SILENCER

Made by
THE MAXIM SILENCER COMPANY
Hartford, Conn.

THE TRUE SPORT FOR THE RED BLOODED AMERICAN
Shooting the small bore rifle equipped with a Maxim Silencer is exhilarating sport

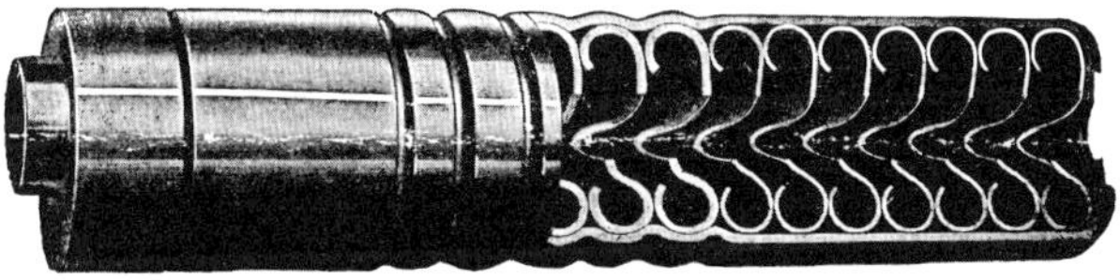

HOW THE SILENCER PERFORMS

The Silencer checks the muzzle blast. Instead of the powder gases being liberated into the air instantaneously when the bullet emerges from the muzzle, as in the ordinary gun, the gases are caught by the Silencer.

They are made to whirl around inside the Silencer. This whirling forces the gas to fly out from the center by centrifugal force leaving a central space, just the same as when water is whirled around in a set bowl, a hole or space, forms in the center. This leaves the space for the bullet to make its passage. The gas cannot pass through this space, until it slows down. This causes it to discharge into the atmosphere gradually. This absolutely prevents report noise and also reduces recoil over two-thirds.

As the hole in the Silencer is much larger than the bullet the latter does not touch anything in passing through and consequently accuracy of flight is just the same whether the Silencer is off or on.

HOW MUCH IT REDUCES THE NOISE

The Silencer absolutely annuls all of the noise of the gun report, as can easily be understood from its construction. It also reduces the recoil over two-thirds. The reason for this is that there is a very strong tendency for the Silencer to be blown off the end of the gun. This tendency amounts to a forward pull on the gun barrel. This forward pull counterbalances part of the backward kick of the gun barrel.

The only noise the Silencer does not control is the noise made out in the air beyond the gun by a high velocity bullet in its flight. This noise is a "crack" like the crack of a whip and not a "whistle" or a "shriek" as is commonly supposed. The noise is made in the wake of the bullet and is caused by the same thing that causes the air to crack when a whiplash is snapped.

This noise cannot be avoided when the bullet velocity is high no matter how quiet we make the gun. Where strictly noiseless bullet flight is wanted in addition to a noiseless shooting gun, modified ammunition has to be used.

Its velocity is 1100 feet per second and a heavy bullet is used so as to keep up the striking energy notwithstanding the reduced velocity. It is very accurate up to 250 yards.

The Maxim Silencer is a practical device for use on **REPEATING** and **SINGLE SHOT RIFLES** and on **SINGLE SHOT TARGET PISTOLS.**

Shot cartridges work perfectly in the Silencer and make noiseless miniature trap shooting possible without disturbance or danger. On single shot target pistols it improves marksmanship to a surprising degree as it avoids nearly all recoil.

Price List of Complete Silencers and Accessories on inside pages.

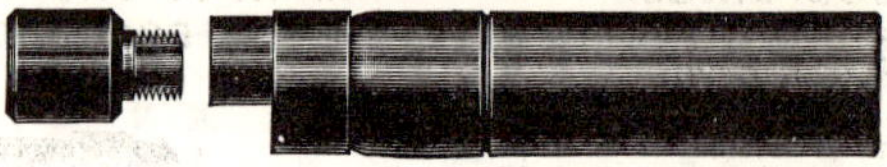

.22 Cal. Silencer and Coupling.

Coupling drives tight on end of barrel. Silencer attaches or detaches by single twist of the wrist.

Dimensions and price including Coupling, Model 1912, 1 in. x 4⅜ in., wt. 5 oz., $5.00.

When ordering, state MAKE and MODEL of rifle to be fitted, and whether barrel is ROUND or OCTAGON, so we can supply proper size coupling.

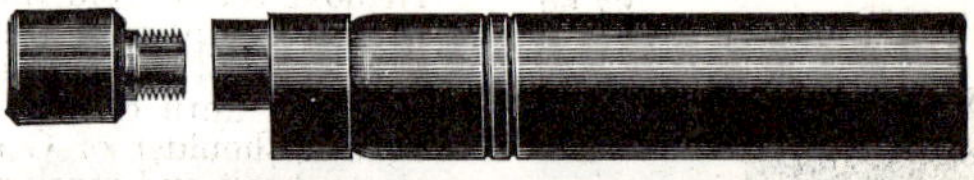

.25-20 Silencer and Coupling.

Coupling drives tight on end of barrel. Silencer attaches or detaches by single twist of the wrist.

Dimensions and Price including Coupling, Model 1912, 1 in. x 5¾ in., wt. 7 oz., $7.00.

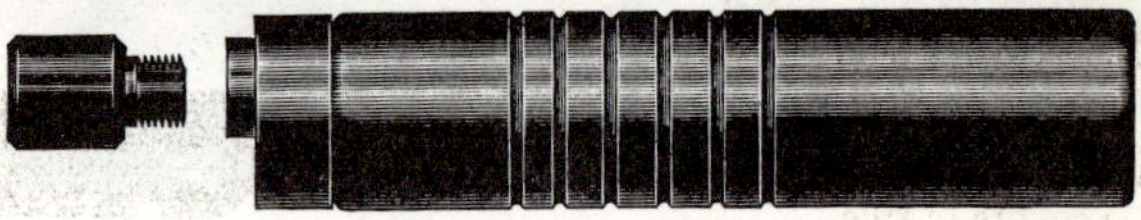

.32-20 Silencer and Coupling.

Coupling drives tight on end of barrel. Silencer attaches or detaches by single twist of the wrist.

Dimensions and Price including Coupling, Model 1909, 1⅜ in. x 6¼ in., wt. 12 oz., $7.00.

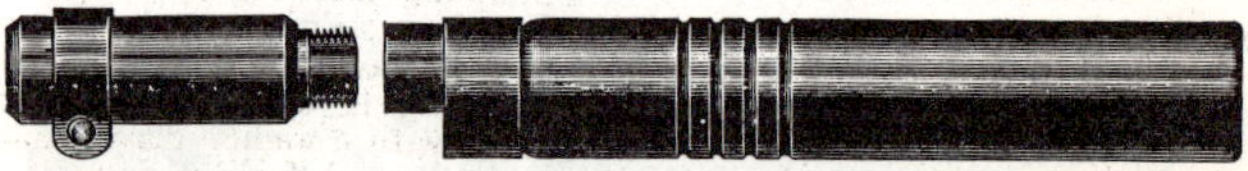

.22 H. P. Silencer and Clamp Type Coupling for Savage H. P. .22 cal. rifle.

Coupling locks on barrel by means of locking clamp. Silencer attaches or detaches by single twist of wrist.

Dimensions and Price including Coupling, Model 1912, 1 in. x. 5¾ in., wt. 7 oz., $8.50.
Price, Silencer only $6.00. Price, Coupling only $2.50.

All calibres larger than above take same price, $9.50 with coupling and $7.00 without coupling

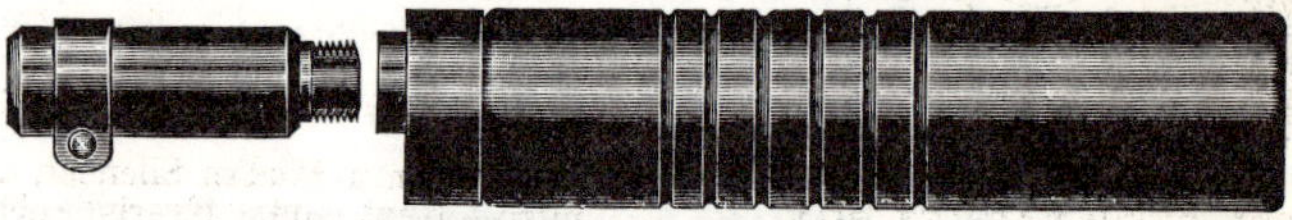

High Power Silencer and Clamp Type Coupling. Any Calibre.

For rifles having front sight MOUNTED UPON BASE built onto barrel. Silencer attaches and detaches by single twist of the wrist. Full length magazine must be shortened to make room for coupling.

Dimensions and Price including Coupling, Model 1909, 1⅜ in. x 6¼ in., wt. 12 oz., $9.50.
Price, Silencer only $7.00. Price, Coupling only $2.50.

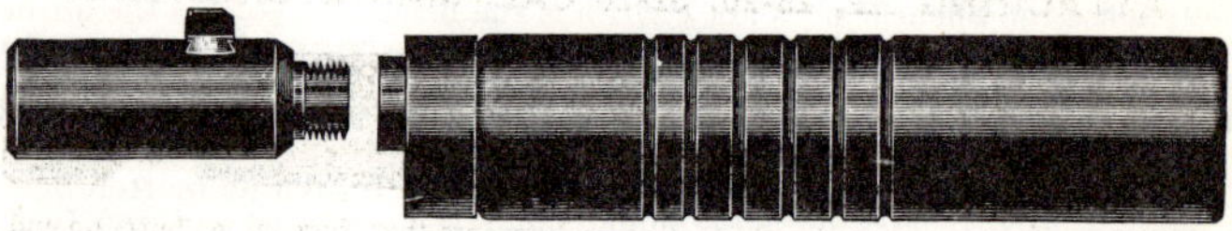

High Power Silencer and Sight Type Coupling. Any Calibre.

For rifles having front sight DOVETAILED INTO BARREL. Coupling drives tight on end of gun barrel after removing front sight from rifle. Silencer attaches and detaches by single twist of the wrist. Full length magazine must be shortened to make room for coupling.

Dimensions and Price including Coupling, Model 1909, 1⅜ in. x 6¼ in., wt. 12 oz., $9.50.
Price, Silencer only $7.00. Price, Coupling only $2.50.

DRIVE-ON COUPLINGS.

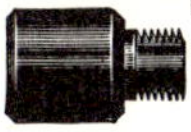

For .22 Cal. Rifles, each $1.00
For .25-20 Cal. Rifles, each 1.00
For .32 Cal. Rim Fire Rifles, each 1.00

SIGHT TYPE COUPLINGS.

For any Calibre Rifle having Sight dove-tailed in barrel, each **$2.50**

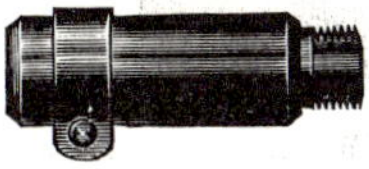

CLAMP TYPE COUPLINGS.

For any Calibre Rifle, each $2.50

When ordering any type of Coupling be sure to mention make, model and calibre of rifle for which desired. Also shape of barrel and length of magazine.

THREAD-PROTECTING CAPS.

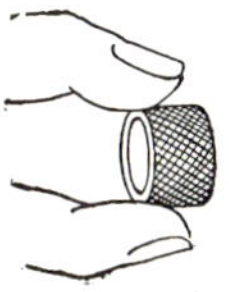

For any Calibre Rifle, each .... 25c.

When ordering Caps mention calibre of rifle.

Threading Barrels to receive Silencers, thereby eliminating the Coupling, each $3.00

Where special work is required in threading the price is based on the time required.

SHIMS.

Shims for Fitting Coupling, per set 25c

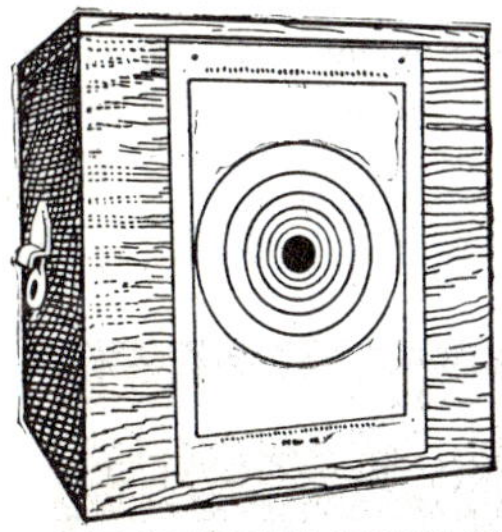

BULLET-STOP TARGET BOX.

A double compartment box made of natural oak. One compartment is to be filled with sand and the other with cotton waste. Just the thing for target practice at home. Will last a lifetime.

Each $4.00

INSTRUCTIONS FOR ATTACHING HI POWER COUPLINGS.

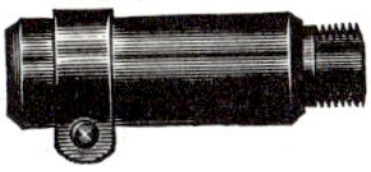

CLAMP TYPE.

First, clean gun barrel and Coupling. Remove screw from the clamp, then turn the clamp upside down on Coupling; this will permit of the Coupling being pushed on to barrel until end of barrel seats firmly against shoulder of Coupling. Now turn clamp back and mark place where groove should be filed in barrel for tightening screw. Remove Coupling and with a rat-tail file, file groove for screw. Now replace Coupling on barrel, insert screw and tighten firmly.

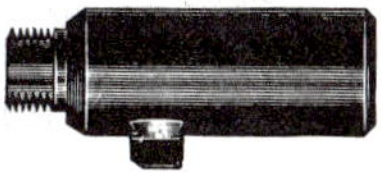

SIGHT TYPE COUPLINGS.

Drive out the sight from the barrel to which the Coupling is to be attached. Coupling includes new sight. Force the Coupling onto the end of barrel, making sure the Coupling is true to alignment after Coupling is on about one-quarter inch. Then with a mallet drive Coupling home until a solid feeling denotes the end of barrel is fitting firmly against shoulder or bottom of Coupling. The Silencer can be attached or detached at will by a half turn. A screw cap is furnished for Coupling end when Silencer is removed.

We recommend the threading of barrels on all Hi Power and octagon barrel rifles.

HOW TO CLEAN THE SILENCER.

To clean a Maxim Silencer, use a good nitro-solvent oil. Nearly all sporting goods stores sell it.

If you cannot secure the oil, soak over night in warm water, rinse thoroughly, and dry quickly on a hot surface to evaporate the water inside. Then oil thoroughly with a good gun oil.

ATTACHING .22, 25-20, 32-20 CAL. MAXIM SILENCERS

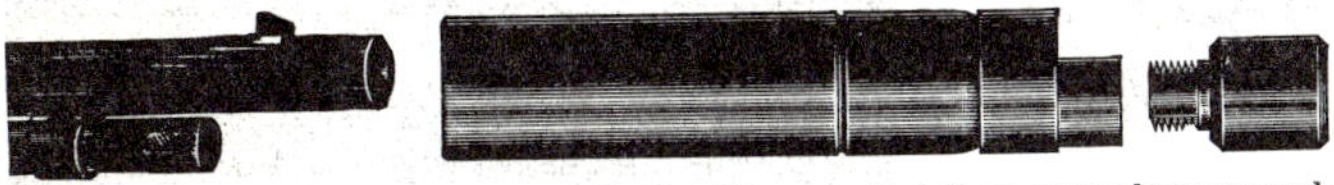

First: Examine end of your barrel and coupling, making sure that there are no burrs on end of barrel or chips in the Coupling.

Second: Remove the spacers which will be found in the Coupling when it is received. Insert as many spacers as are necessary to make Coupling just start over extreme end of barrel. Special care should be taken to see that the thin shims do not get jammed or folded over. This will cause the Coupling to be deflected and out of true alignment. It is only necessary to start the Coupling on the barrel about ⅛ inch.

Third: Now twist the Silencer and Coupling around on the barrel until the groove on the under side of Silencer comes in the center of the bottom side of the rifle or magazine tube. Rap the end of Silencer with a soft piece of wood just enough so that the Coupling will not move while you are removing the Silencer. With a mallet, drive home the Coupling until it gives forth a solid sound and feeling.

Fourth: If your rifle is a Winchester or Marlin with tubular magazine, the Silencer is simply turned bottom side up which allows the tube to be withdrawn to load the rifle.

Threading gun barrels for attaching Maxim Silencers is strongly advised for all heavy calibre rifles. Any good gunsmith can do this work by strictly following instructions below.

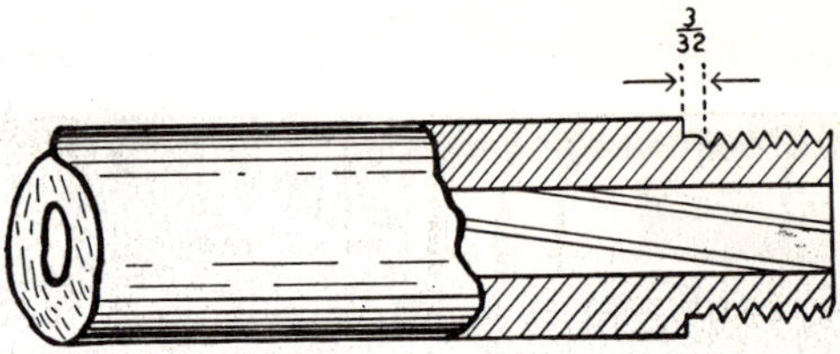

INSTRUCTIONS FOR THREADING RIFLE BARREL.

1. The following are the regular threads used on the various calibre Silencers.

.22 cal. ½ inch, .20 thread. .30 cal. to .35 cal. inc. 9-16 in., 20 thread
.25 cal. ½ inch, .20 thread .38 cal. and .40 cal. .614 in., 20 thread.
.44 cal. and .45 cal. 11-16 inch, 20 thread.

2. The neck of the Silencer will be found to contain a counterbored portion ⅛ inch deep, and the outside diameter of the thread. THERE MUST POSITIVELY BE AN UNTHREADED SPACE ON THE BARREL TO FIT THIS COUNTERBORE. This is very important, as it is this fit that insures the correct alignment of the Silencer on the gun barrel. The illustration shows this unthreaded portion.

3. The fit of the threads should be just loose enough to enable the Silencer to be screwed on without difficulty by hand. The shoulder on the gun barrel should just bring the Silencer so that the groove in the bottom points straight down.

4. The thread in the Silencer will be found to be interrupted. If the thread on the gun barrel is interrupted on opposite corners to correspond, the Silencer can be dropped into place and fastened by a single quarter turn.

5. If the bore in the Silencer aligns with the bore of the gun barrel, the bullet cannot possibly touch the Silencer after it leaves the gun barrel. The bore of the Silencer is 1-16 inch larger than the diameter of the bullet and ample clearance is therefore given.

MODEL 1912 — .22 CAL. SILENCERS WITH COUPLINGS.

In the table below will be found a list of the different makes of .22 calibre rifles which can be fitted with a Maxim Silencer by the type of Coupling indicated in the margin preceding the name of the rifle.

Coupling	RIFLE	Model	Barrel	Coupling	RIFLE	Model	Barrel
A	Winchester	1902	Round	4-A	Stevens Reliable Pistol	42	Round
A	Winchester Automatic	1903	Round	4-A	Stevens Diamond Pistol	43	Round
A	Winchester Thmb. Trgr.		Round	2-B	Stevens Marksman	12	Round
A-Flat	Winchester	1906	Round	2-B	Stevens Ideal	044½	Round
B	Winchester	1890	Oct.	2-B	Stevens Ideal	56	Round
C	Winchester	1904	Round	2-B Short	Stevens Marksman	12	Round
E	Winchester Single Shot	No. 1	Oct.	2-B Short	Stevens Ideal	40	Round
I	Winchester Single Shot	No. 3	Oct.	2-B Short	Stevens Vernier	40½	Round
I	Winchester Single Shot	No. 3	Round	2-B Short	Stevens Off-Hand Pistol	35	Round
J	Winchester Single Shot	No. 4	Round	C	Stevens	27	Oct.
M	Winchester Single Shot	No. 1	Round	E	Stevens Ideal	44	Oct.
N	Winchester Single Shot	No. 1	Round	E	Stevens Ideal	44½	Oct.
O	Winchester Single Shot	No. 3	Oct.	E	Stevens Armory	414	Round
A	Remington, No. 1	12	Round	E	Stevens Semi Military	404	Round
A	Remington Automatic	16	Round	E	Stevens Ideal Range	45	Round
B	Remington	4	Oct.	E	Stevens Gallery	80	
B	Remington, 2-B	12	Oct.	K	Stevens Favorite	17	Round
B	Remington, 3C	12	Oct.	F. S. Flat	Stevens Visible	70	Round
2-B	Remington	6	Round	A Short	Savage	1909	Round
F	Remington Single Shot	R. W.	Oct.	A Short	Savage Automatic	1912	Round
A	Marlin	25	Round	B	Savage	1903	Oct.
A-Flat	Marlin	1892	Round	B	Savage	1905	Round
A-Flat	Marlin	1897	Round	B	Savage	1914	Oct.
2-A	Marlin	18	Round	2-B	Savage	1904	Round
2-A	Marlin	18	Oct.	2-B	Savage	1911	Round
2-B	Marlin Half Magazine	1897	Oct.	A	Hopkins & Allen	722	Round
2-B Flat	Marlin Full Magazine	1897	Oct.	2-B	Hopkins & Allen	822	Round
2-B	Marlin	32	Oct.	C Short	Hopkins & Allen, Jr.	922	Round
2-B-Flat	Marlin	1892	Oct.	E	Hopkins & Allen	3922	Oct.
C	Marlin Half Magazine	20	Oct.	H	Hopkins & Allen	2922	Oct.
C-Flat	Marlin Full Magazine	20	Oct.	I	Hopkins & Allen Hvy-wt		Oct.
K	Marlin	29	Round	A	Meriden	J-10	Round
A	Stevens Little Krag		Round	B	Meriden	K-15	Oct.
A-Short	Stevens Crackshot	16	Round	2-A	Quackenbush		Round
A-Short	Stevens Little Scout	18	Round	2-B Short	Colt Repeater		Round
2-A	Stevens Maynard, Jr.	15	Round	F	Batavia Automatic		
2-A	Stevens Maynard, Jr.	15½	Round	G	Winder Musket		

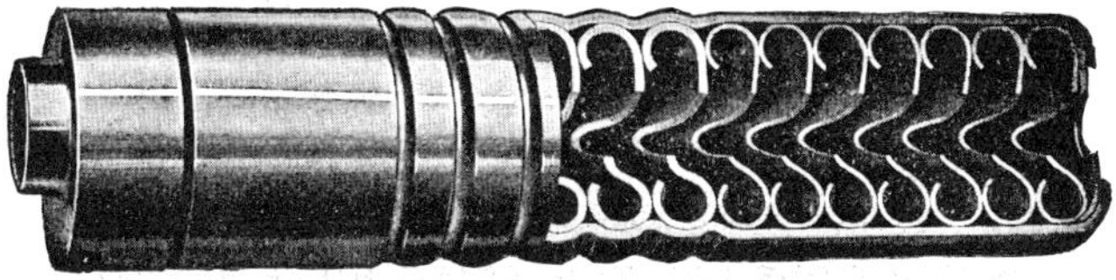

MAXIM SILENCER.

IT WILL PAY YOU TO KNOW ABOUT IT.

PRESERVE THIS SHEET FOR REFERENCE.

HOW IT WORKS.

The Silencer checks the muzzle blast. Instead of the powder gases being liberated into the air instantaneously when the bullet emerges from the muzzle, as in the ordinary gun, the gases are caught by the Silencer.

They are made to whirl around inside the Silencer. This whirling forces the gas to fly out from the center by centrifugal force leaving a central space, just the same as when water is whirled around in a set bowl, a hole or space forms in the center. This leaves the space for the bullet to make its passage. The gas cannot pass through this space, until it slows down. This causes it to discharge into the atmosphere gradually. This absolutely prevents report noise and also reduces recoil over two-thirds.

As the hole in the Silencer is much larger than the bullet, the latter does not touch anything in passing through and consequently accuracy of flight is just the same whether the Silencer is off or on.

THE DIFFERENT SIZES.

Silencers are furnished for every calibre rifle from the .22 up to the .45 inclusive. The only rifles excepted from this list are the large calibre Winchester and Remington auto-loaders. For these we only furnish Silencers by special arrangement.

The .22 auto-loading and all calibres of the Standard auto-loading rifles have regular Silencers especially adapted for them.

Single shot rifles, repeating rifles, carbines and single shot target pistols of any make or calibre can be fitted with Silencer.

Shot-gun Silencers are not yet ready for distribution. Shot loaded cartridges can be used perfectly in rifles fitted with regular rifle Silencers.

Revolvers and automatic pistols are not adapted to be silenced and therefore cannot be fitted.

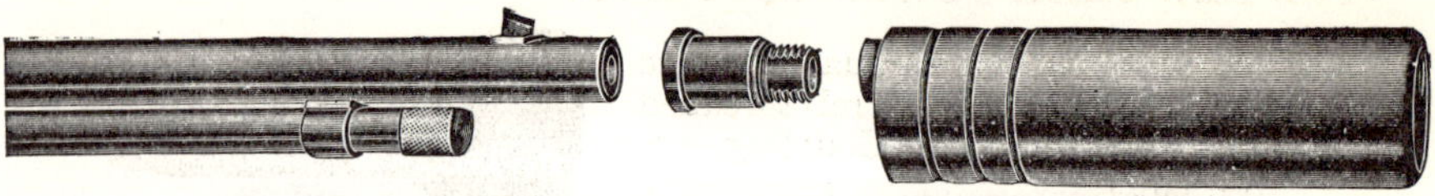

Rifle Coupling Silencer

HOW THE .22 CAL. IS ATTACHED.

For all .22 cal. rifles or single shot pistols the Silencer can be immediately attached by means of a Coupling which drives onto the end of the barrel forward of the front sight. The cut shows this. The Coupling is a small thimble having the threads by means of which the Silencer is screwed on. Spacers accompany it so that a tight driving fit can be obtained no matter what size the gun barrel is. Various sizes Couplings are furnished **(See List of .22 Calibre Couplings)** so that only a few spacers are necessary. The round Coupling drives onto either octagon or round barrels perfectly. A small thimble covers the threads of the Coupling when the Silencer is off and gives the barrel a finished appearance. A .22 outfit consists of a Silencer, one Coupling of any size desired, several spacers and a Barrel Sleeve or Thimble and full instructions.

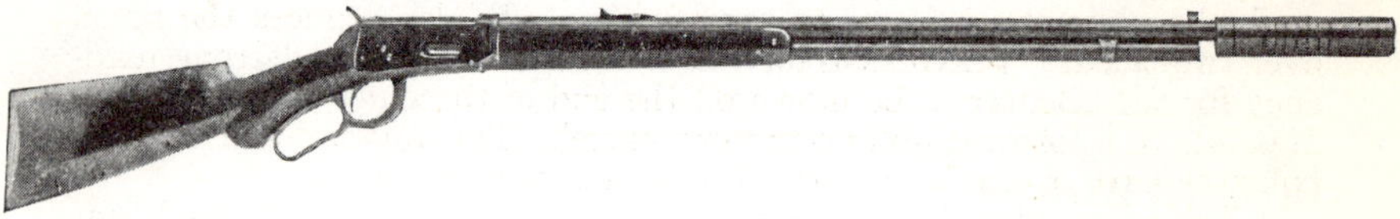

HOW THE 25-20 CAL. AND 32-20 CAL. IS ATTACHED.

These are equipped in the same way as the .22 except that to positively lock the Coupling on the barrel a locking pin is used in the Coupling. (See Cut.) A small groove has to be filed in the bottom of the barrel forward of the front sight to pass this locking pin. Only round barrels can be fitted this way. Octagon barrels must be turned down round forward of the front sight, or what is better still must be threaded.

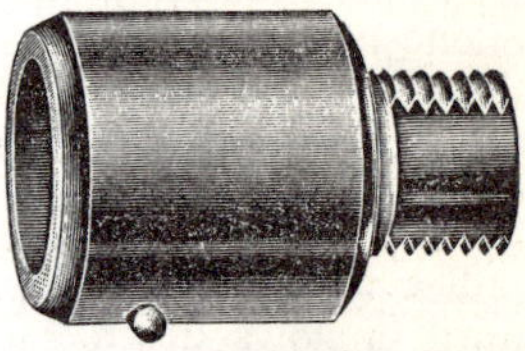

.25-20 and .32-20 Coupling.

HOW ALL OTHER CALIBRES ARE ATTACHED.

The best way to attach any Silencer is to have the gun barrel threaded forward of the front sight. As is apparent, this entirely avoids a Coupling with its cost, weight and appearance. The barrel sleeve covers these threads when the Silencer is off and makes the barrel thus fitted look the same as a regular barrel. It makes a first class job in every way. By removing the sleeve, the Silencer can be attached by a single twist of the wrist. It is not a difficult job for any machinist to do the threading if the instructions accompanying each Silencer are observed. The threading does not injure the gun barrel, or the shooting, or the strength of the gun in any way whatsoever. (See Cut.) In ordering new guns, the dealer should always remember to order part of them to come threaded to take Silencer. Then Silencer can be attached at any time. It also helps sell the gun.

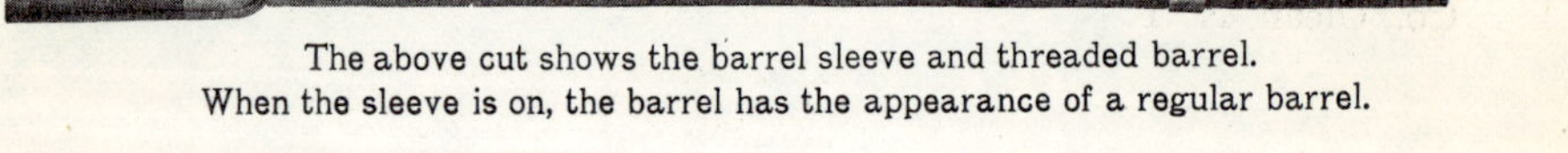

The above cut shows the barrel sleeve and threaded barrel.
When the sleeve is on, the barrel has the appearance of a regular barrel.

THE NEW HIGH POWER COUPLING.

For those who are situated where they cannot have the barrel threaded we furnish a High Power Coupling. (See cut.)

This Coupling consists of a thimble carrying the threads for attaching the Silencer, a taper locking grip which encircles the barrel and a hexagon locking nut for tightening the grip. It can be attached to the round barrel rifles only. Each make and model of rifle has its own special Coupling and in ordering this information must always be given. The front sight has to be driven out to pass the locking nut onto the barrel. The sight is driven back into its place after the locking nut is on. The barrel sleeve covers the threads when the Silencer is off and gives a finished appearance.

Where rifles or carbines have the tubular magazine running all the way out to the end of the barrel, it is necessary to shorten the magazine two and one-half inches. This is easily done on all rifles, and does not seriously reduce the magazine capacity.

HOW MUCH IT REDUCES THE NOISE.

The Silencer absolutely annuls all of the noise of the report, as can easily be understood from its construction. It also reduces the recoil over two-thirds. The reason for this is that there is a very strong tendency for the Silencer to be blown off the end of the gun. This tendency amounts to a forward pull on the gun barrel. This forward pull counterbalances part of the backward kick of the gun barrel.

The only noise the Silencer does not control is the noise made out in the air beyond the gun by a high velocity bullet in its flight. This noise is a "crack" like the crack of a whip and not a "whistle" or a "shriek" as is commonly supposed. The noise is made in the wake of the bullet and is caused by the same thing that causes the air to crack when a whiplash is snapped.

This noise cannot be avoided when the bullet velocity is high no matter how quiet we make the gun. Where strictly noiseless bullet flight is wanted in addition to a noiseless shooting gun, modified ammunition has to be used. We furnish this at the same price as regular ammunition. Its velocity is 1100 feet per second and a heavy bullet is used so as to keep up the striking energy notwithstanding the reduced velocity. It is very accurate up to 350 yards. We furnish it in all calibres in which center-fire cartridges are used.

.22 LONG AND .22 W. R. F. CARTRIDGES.

Some times a .22 Long and a .22 W. R. F. cartridge will be found which has a bullet velocity high enough to make the bullet noise. .22 Short and .22 Long rifle of the regular velocities are always quiet. There is no way to stop the noise of the .22 Long and the .22 W. R. F. bullet. The best thing to do is to try a different make of ammunition until you get one which gives quiet bullet flight.

BLACK OR SEMI-SMOKELESS POWDERS.

With smokeless powder the Silencer never needs any more cleaning than to jar it to shake out any unburned powder grains which may collect. With black or semi-smokeless it must be cleaned. One way is to use hot water, placing the Silencer under the hot water faucet in a sink, and letting the water flow through. In half an hour it will be cleaned. If desired a quick and very thorough cleaning may be accomplished in a few minutes by using the Solvents made expressly for this purpose by Frank A. Hoppe, Philadelphia, Pa., or the Tri-Products Co., Olean, N. Y.

.22 AUTOMATIC AMMUNITION.

Some times an ammunition will be found which will clog the breech action of the .22 automatic Winchester rifle with unburned grains of powder. All makes of ammunition do not do this. If trouble is experienced try another make and the trouble will be entirely overcome.

ACCURATE SHOOTING.

The hole in the Silencer is considerably larger than the diameter of the bullet. If the Silencer is any where nearly in line with the barrel, the bullet cannot touch in passing through. If, however, the alignment is so bad that all the clearance is used up and the bullet touches, its flight will be disturbed and it will not shoot accurately. A light touch too small to be noticed is enough. To correct it, the easiest way is to file with a round file where the bullet is touching and get the necessary clearance. This does not injure the Silencer, as the size of the hole is not important.

DIFFERENT SIGHTING WITH SILENCER.

The addition of the weight of the Silencer changes the vibration of the barrel in a high power rifle. This causes a change in the sighting when the Silencer is used. Usually the sight has to be moved slightly to the right, or if the rifle has windage adjustment on the rear sight correction can be made here.

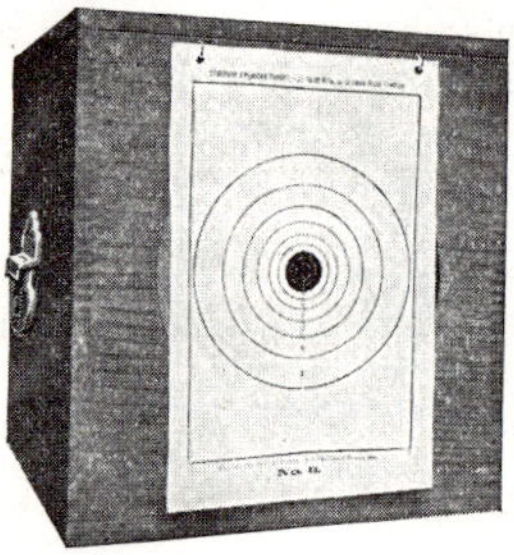

BULLET STOP TARGET BOX.

A real novelty for which there is a large demand is our Bullet Stop Target Box. It is a double compartment box, the larger compartment being filled with dry sand and the smaller with cotton waste. It stops the bullet of even a powerful rifle and does so without noise. The double compartment feature and the waste prevents sand leaking out even after the front of the box is riddled with repeated shooting. The front cloth is easily replaced. The box will last forever. It has natural oak finish and can be set up any where, indoors or out.

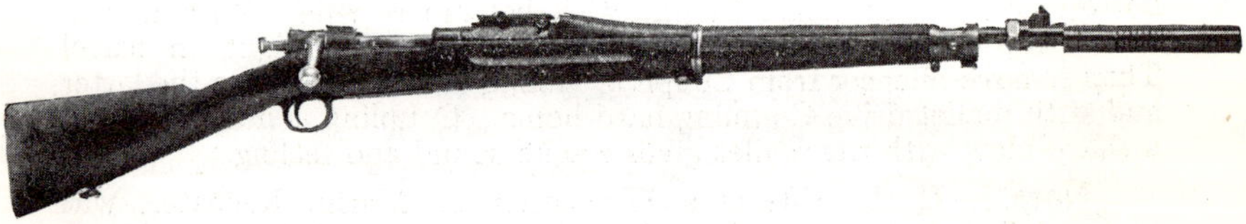

MODEL 15 U. S. GOV. SILENCER.

This is the Silencer issued by the War Department to the National Guard. It needs no coupling and can be attached to the U. S. Government Springfield rifle immediately, by simply removing the front sight blade. It is sold to private owners of the Springfield rifle and also to individual members of the National Guard.

PRICES.

	Net Retail
.22 calibre Silencer outfit complete including Coupling, and Barrel Sleeve with full instructions for attaching	$5.00
Extra .22 cal. Coupling only	1.00
Extra .22 cal. Barrel Sleeve	.25
.25-20 cal. Silencer outfit complete including Coupling and Barrel Sleeve with full instructions for attaching	7.00
Extra .25 cal. Coupling only	1.00
Extra .25 cal. Barrel Sleeve	.25
.32-20 Cal. Silencer outfit complete including Coupling and Barrel Sleeve with full instructions for attaching	7.00
Extra .32 cal. Coupling only	1.00
Extra .32 cal. Barrel Sleeve	.25
High Power Silencer and Barrel Sleeve only	7.00
High Power Coupling only, any cal. with full instructions for attaching	2.50
Official U. S. Gov. Silencer for Gov. Springfield Rifle	8.50
Patent Bullet Stop Target Box for Indoor Shooting	3.50
Adapting barrels to receive Silencer and furnishing Barrel Sleeve	2.00

TERMS.

All deliveries f. o. b. Hartford, Conn., and all Silencers are sold by us and licensed only upon the express condition that there shall be no deviation from the prices here given without our written consent.

ORDER THROUGH YOUR REGULAR JOBBER.

INSTRUCTIONS FOR EQUIPPING A .22 CAL. RIFLE.

First. — Examine end of barrel and Coupling making sure they are clean and free from metal filings.

Second. — Remove the Spacers which will be found in the Coupling when it is received. Insert as many spacers as are necessary to make Coupling just start over extreme end of rifle barrel. Special care should be taken to see that the thin shims do not fold over. It should only be possible to start Coupling onto barrel about one eighth inch.

Third. — Twist Coupling and Silencer around until groove in bottom of Silencer comes in line with bottom of rifle. With a wood mallet rap the end of the Silencer until coupling is tight on barrel. Then remove Silencer from Coupling, taking care not to move the latter, and with mallet drive Coupling hard home. Coupling is not home until a sharp blow with the mallet gives a solid sound and feeling.

Note. — If the rifle is a Winchester or Marlin Repeater, with underneath tubular magazine, the Silencer is turned bottom side up when the magazine is to be filled (see illustration).

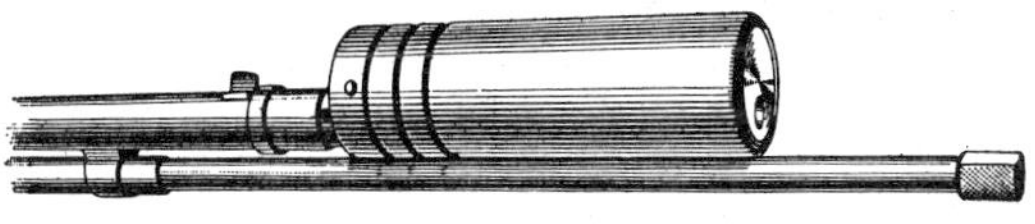

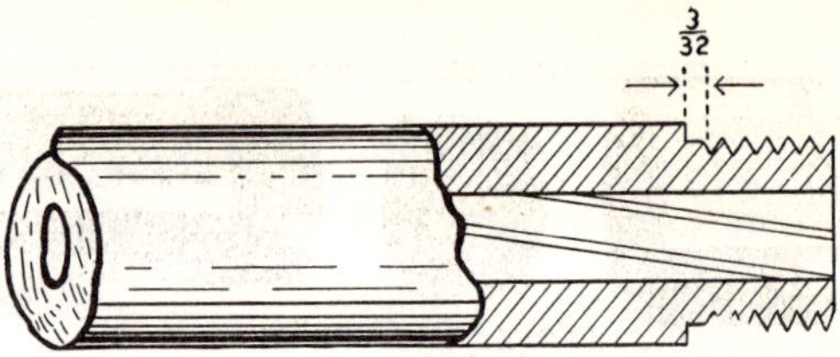

INSTRUCTIONS FOR THREADING RIFLE BARREL.

1. The following are the regular threads used on the various calibre Silencers:
   .22 cal. ½″ 20 thread.
   .25 cal. $\frac{9}{16}$″, 20 thread.
   .30 cal. to .35 cal. inclusive $\frac{9}{16}$″, 20 thread.
   .38 cal. and .40 cal. .614″, 20 thread.
   .44 cal. and .45 cal. $\frac{11}{16}$″, 20 thread.
2. The neck of the Silencer will be found to contain a counterbored portion ⅛″ deep, and the outside diameter of the thread. There should be an unthreaded space on the barrel to fit this counterbore. This is very important as it is this fit that insures the correct alignment of the Silencer on the gun barrel. The illustration shows this unthreaded portion.
3. The fit of the threads should be just loose enough to enable the Silencer to be screwed on without difficulty by hand. The shoulder on the gun barrel should just bring the Silencer so that the groove in the bottom points straight down.
4. The thread in the Silencer will be found to be interrupted. If the thread on the gun barrel is interrupted on opposite corners to correspond, the Silencer can be dropped into place and fastened by a single quarter turn.
5. If the bore in the Silencer aligns with the bore of the gun barrel, the bullet cannot possibly touch the Silencer after it leaves the gun barrel. The bore of the Silencer is $\frac{1}{16}$″ larger than the diameter of the bullet and ample clearance is therefore given.

INSTRUCTIONS FOR ATTACHING HIGH POWER COUPLINGS.

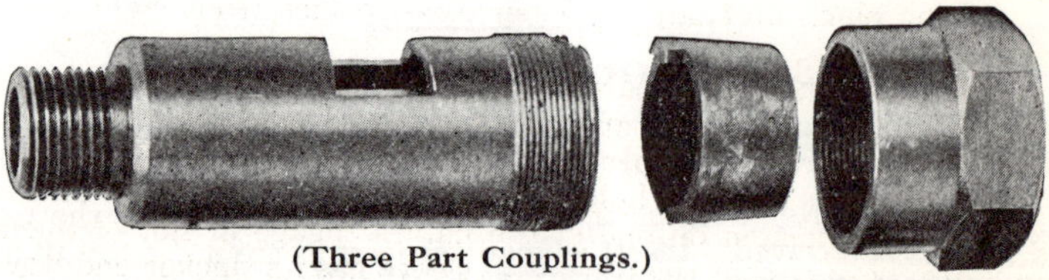

(Three Part Couplings.)

The coupling consists of three pieces: — The Body, the Split Sleeve, and the Lock Nut.

1. Dis-assemble the Coupling by unscrewing the lock nut and removing from it the split sleeve.
2. To attach to rifle: — Remove the front sight and pass the lock nut over the barrel, having the threaded end point toward the muzzle.
3. Pass the split sleeve over the barrel, taking care to see that the little projection on the sleeve points towards the muzzle.
4. Pass the body of Coupling onto the end of the gun barrel and force on until it bottoms firmly. Rap it with a piece of hard wood.
5. Engage the projection on the split sleeve in the slot of coupling body.
6. Screw up the lock nut as tight as possible by means of an 8″ wrench, TAKING PARTICULAR CARE TO SEE THAT PROJECTION ON SPLIT SLEEVE DOES NOT DISENGAGE FROM SLOT IN COUPLING. Use a little oil or grease on surface of split sleeve and threads.
7. Replace the front sight.

(Four Part Couplings.)

The Coupling consists of four pieces: — The Body, the Split Sleeve, the Locking Nut and the Nut Case.

1. Dis-assemble the Coupling by unscrewing the locking nut and removing from the nut the split sleeve, and separating the nut from its case.
2. To attach to rifle, first take out the front sight. Slip the nut case over the end of barrel with the marked end towards the rear.
3. Slip the locking nut over the barrel and into its nut case matching up the numbers.
4. Slip the split sleeve over the barrel, taking care to see that the little projection on this sleeve points TOWARDS THE MUZZLE.
5. Slip the body of Coupling over the end of barrel and force down till it bottoms firmly.
6. Bring the projection on split sleeve to a position so that it WILL ENGAGE in slot of Coupling body.
7. Screw up the locking nut as tight as possible by means of a 6″ or 8″ wrench, taking care to see that the split sleeve does not disengage from slot in body of Coupling. Use a little oil or grease on the outer surface of the taper sleeve and threads.
8. Replace the front sight.

INSTRUCTIONS FOR ATTACHING MODEL 15 SILENCER TO U. S. SPRINGFIELD RIFLE.

The Coupling consists of five pieces:—The Body, the Two Halves of Taper Sleeve, with Spring for holding the two latter in place and the Nut.

1. Dis-assemble by unscrewing the Nut and removing the halves of the Taper Sleeve.
2. Remove small screw which holds front sight dovetail in place, and remove front sight. This permits Nut to be slipped over sight block and onto gun barrel. Replace front sight dovetail and screw.
3. Pass Silencer on to gun barrel and force home until end of slot in Silencer is within $\frac{1}{16}$″ of front sight block on barrel.
4. Place two halves of Taper Sleeve around barrel and snap Spring into groove so as to hold the halves in place. Be sure and have small projection on one of the halves engage in slot of Silencer. Also be sure that the other half sleeve has its thicker end toward muzzle of rifle.
5. Screw up Nut onto Silencer as tightly as possible by means of a wrench. Use a little oil on outer surface of Taper Sleeve and on threads.

LIST OF .22 CAL. COUPLINGS.

Rifles.	Name of Coupling.	Inside Diam.
Stevens "Reliable" Model 42. Stevens "Diamond" Model 43.	4A	.431 (approx. $\frac{7}{16}$)
Stevens "Little Scout" Model 14 Stevens "Maynard Jr." Model 15. Quackenbush.	2A	.515 (½ full)
Stevens "Little Krag." Marlin Model 20, Oct. bbl. Marlin Model 25, Rd. Bbl. Winchester Model 1902, S. S. Rd. Bbl. Winchester Model 1903, Auto., Rd. Bbl. Winchester Thumb Trigger, S. S.	A	.586 (approx. $\frac{19}{32}$)

Rifles.	Name of Coupling.	Inside Diam.
Savage Model 1909, Rd. Bbl. Hopkins & Allen, Model 22, Rd. Bbl. Remington, Model R. W. S. S., Rd. Bbl. Remington, Model 12, No. 1, Rd. Bbl.	**A Short**	.586 (approx. 19/32)
Winchester, Model 1906, Rd. Bbl. Marlin Model 1892, Rd. Bbl. Marlin, Model 1897, Rd. Bbl.	**A Short, Flatted**	.586 (approx. 17/32)
Stevens "Ideal Ladies" Model, No. 56. Stevens Model 044½. Savage Model 1904, Rd. Bbl. Savage Model 1910, Rd. Bbl. Hopkins & Allen, Model 822 and 832, Round Barrel. Remington, Model 6, S. S., Rd. Bbl.	**2B**	.615 (⅝ scant)
Stevens New Model No. 40. Stevens New Model Vernier, No. 40½. Stevens "Crack Shot," No. 16. Stevens' 'Off Hand," No. 35. Colt Repeater, Oct.	**2B Short**	.615 (⅝ scant)
Savage Model 1911. Remington R. W. S. S., Oct. Bbl. Batavia Automatic. Stevens Gallery, No. 80, Rep. Marlin Model 1892, Oct. Bbl. Marlin Model 1897, Oct. Bbl.	**F**	.629 (⅝ full)
Stevens Visible Loading Model 70.	**F Short, Flatted**	.629 (⅝ full)
Stevens "Favorite," No. 17. Hopkins & Allen Model 922, Rd. Bbl. Hopkins & Allen, Model 4922, Rep. Nos. 2 and 3, Oct. Bbl.	**B Short**	.679 (11/16 scant)
Remington Model 12, No. 2 and 3, Oct. Bbl. Winchester Model 1890, Oct. Bbl. Savage Model 1903, Oct. Bbl.	**B**	.679 (11/16 scant)
Savage Model 1905, Rd. Bbl. Winchester Model 1904, S. S., Rd. Bbl.	**C**	.692 (11/16 full)
Remington Model 4, S. S., Oct. Bbl.	**C Short**	.692 (11/16 full)
"Winder Musket."	**G**	.710 (23/32 full)
Hopkins & Allen, Model 1922, Oct. Bbl. Winchester S. S., No. 1 Barrel, 24-inch, Rd. and Oct. Winchester, S. S., No. 1 Barrel, 26-inch, Rd. and Oct.	**H**	.785 (approx. 25/32)
Stevens, Model, No. 44. Stevens "Ideal," No. 44½. Stevens "Ideal Range," No. 45. Hopkins & Allen, Light Weight Schuetzen.	**E**	.828 (13/16 full)
Special.	**D**	.880 ⅞ full)
Hopkins & Allen, Heavy Weight Schuetzen. Winchester S. S., No. 3 Barrel, 24-inch and 26-inch, Rd. and Oct.	**I**	.937 (15/16)
Winchester S. S., No. 4 Barrel, 24-inch and 26-inch, Rd. and Oct.	**J**	1.080 (1 1/16 full)

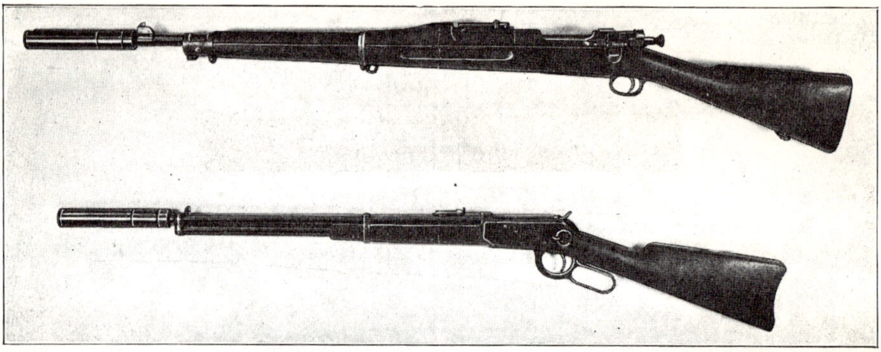

Springfield .30 Cal. Rifle with Maxim Silencer.
Winchester .30-30 Carbine with Maxim Silencer.

HOW THE SILENCER OPERATES

The Silencer checks the muzzle blast. Instead of the powder gases being liberated into the air instantaneously when the bullet emerges from the muzzle, they are caught by the Silencer, and allowed to escape gradually.

This is accomplished by the SILENCING CHAMBERS, which are so constructed as to offer a very high resistance to a rapid flow of gas through them, but very little resistance to a slow flow of gas.

The hole in the Silencer is much larger than the bullet and the latter does not touch anything in passing through, and consequently accuracy of flight and velocity are just the same with or without the Silencer.

HOW MUCH IT REDUCES THE NOISE

The Silencer annuls almost all the noise of the gun report. On high power rifles it also reduces the recoil over two-thirds. The reason for this is that there is a very strong tendency for the Silencer to be blown off the end of the gun. This tendency amounts to a forward pull on the gun barrel. This forward pull counterbalances part of the backward kick.

The only noise the Silencer does not control is the noise made out in the air beyond the gun by a bullet in its flight. This noise is a "crack" like the crack of a whip and not a "whistle" or a "shriek" as is commonly supposed. The noise is made by the same thing that causes the air to crack when a whiplash is snapped.

This noise cannot be avoided when the bullet velocity exceeds 1,075 feet per second, no matter how quiet we make the gun. .22 Short and .22 Long Rifle ammunition is below this, and flies quietly. .22 Long and .22 W. R. F. and all high power ammunition are above the critical velocity and usually give bullet flight noise.

Shot cartridges up to .45 calibre work perfectly in the Silencer and make noiseless miniature trap shooting possible without disturbance or danger. On single shot target pistols it improves marksmanship to a surprising degree as it avoids nearly all recoil, or "flip."

The Maxim Silencer will not operate on revolvers. No revolver can ever be silenced. If you stop the noise at the muzzle, it makes its escape at the gap between the cylinder and the barrel. Every revolver must have this gap.

THE MAXIM SILENCER IS A PRACTICAL DEVICE FOR USE ON ALL REPEATING, AUTOMATIC, SINGLE SHOT RIFLES AND SINGLE SHOT TARGET PISTOLS.

It is made in all calibres from .22 to .45.

PRICE LIST

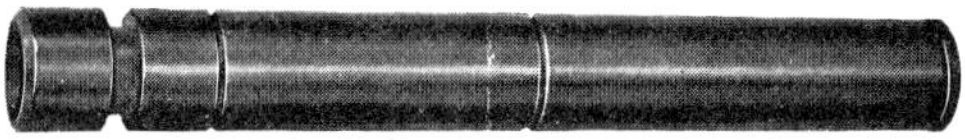

Model 1923 .22 Cal. Silencer and Coupling. Can be taken apart.
Size: Diam. ⅞ in., length 6⅝ in., weight 5½ oz.

Price: Silencer with any size Coupling, Shims, Thread Cap and Assembling Tool.................... $7.50
Silencer and Assembling Tool.. 6.50

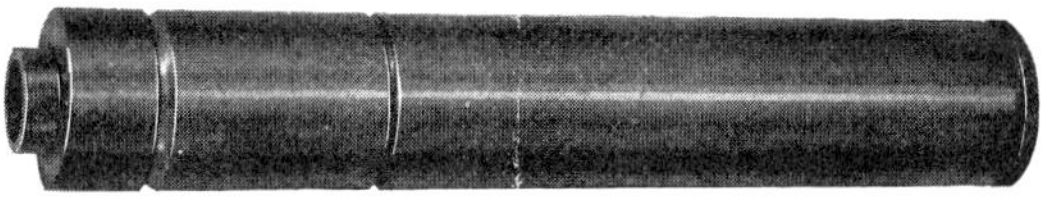

Model 1921 High Power Silencer. Can be taken apart.
Size: Diam. 1⅛ in., length 6⅞ in., weight 10½ oz.
Made in calibres from .22 H. P. to .32 inclusive. Gun barrel must be threaded to take Silencer.

Price: Silencer, Thread Cap and Assembling Tool.................................... $10.50
Our charge for threading gun barrel.. 3.00

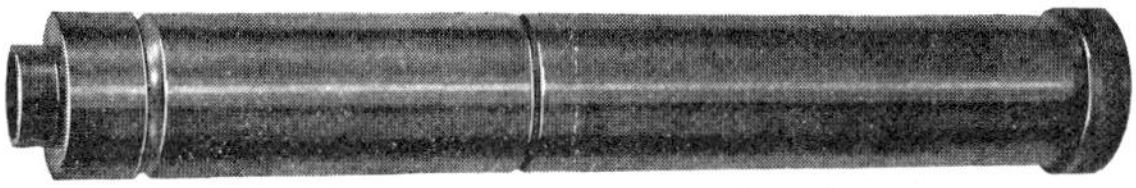

Model 1924 High Power Silencer. Can be taken apart.
Size: Diam. 1⅛ in., length 8 in., weight 12¼ oz.
Made for specially high power .30 and .32 cal. rifles.

Price: Silencer, Thread Cap and Assembling Tool.................................... $12.50
Our charge for threading gun barrel.. 3.00

PARTS

.22 cal. Coupling, Thread Protecting Cap and Shims	$1.50
Thread Protecting Cap, .22 cal. up to .32 cal	.40
Thread Protecting Cap, above .32 cal	.45
Assembling Tool	.50
One set of .22 cal. Shims	.30
Model 1923 .22 cal. Silencer Body	4.00
Model 1923 .22 cal. Silencing Chambers, each	.10
Model 1923 .22 cal. Silencer End Cap	1.00
Model 1923 .22 cal. Plain Washer	.20
Model 1921 High Power Silencer Body	6.00
Model 1921 High Power Silencing Chambers, each	.20
Model 1921 High Power Silencer End Cap	1.50
Model 1921 High Power Plain Washer	.20
Model 1924 High Power Silencer Body	7.00
Model 1924 High Power Silencing Chambers, each	.20
Model 1924 High Power Silencer End Cap	2.00
Model 1924 High Power Plain Washer	.20
Charge for threading gun barrels when sent to us	3.00
Charge for shortening magazine tubes when done by us	1.50

INFORMATION

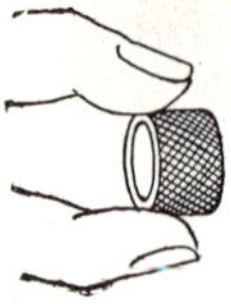

THREAD PROTECTING CAPS

These caps screw onto the threaded portion of a coupling or a gun barrel, and protect the threads and give the barrel a neat appearance when the Silencer is off.

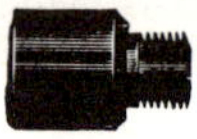

DRIVE ON COUPLINGS

These couplings furnish the means for attaching Silencers to .22 calibre rim fire rifle barrels. Thread Protecting Cap is furnished with each coupling.

SHIMS

Shims are used inside the coupling so as to secure the proper tight fit between the gun barrel and the coupling. They are madé of tempered steel .002″ and .005″ thickness. One set of shims consists of three thick shims and two thin shims.

THREADED TYPE COUPLINGS

This coupling is furnished for attaching Silencer to small .22 cal. rifle and pistol barrels.

ASSEMBLING TOOL

This tool is necessary in order to take the Silencers apart. Without it you are pretty sure to spoil some of the silencing chambers.

DIRECTIONS FOR ASSEMBLING SILENCERS

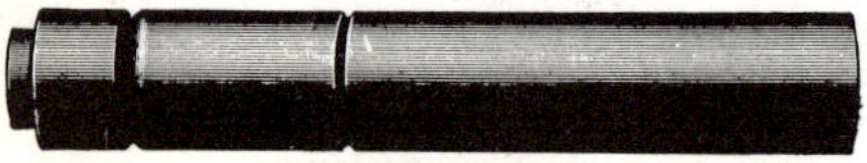

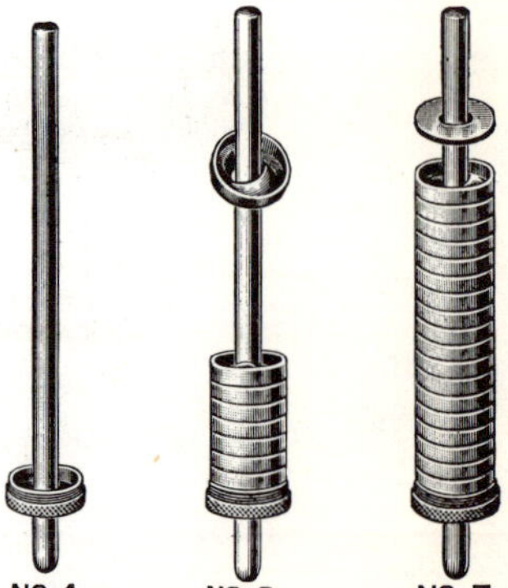

1. Take Tool in left hand, holding long end up. Place Knurled End Cap on Tool, with CUPPED SIDE UP as in Fig. 1.

2. Place eighteen Silencing Chambers on Tool, being sure the CUPPED ENDS ARE UP, and that the Chambers match each other — lump over lump, or hollow over hollow, as in Fig. 2.

3. Place Plain Washer on Tool as in Fig. 3. You are now ready to place the whole thing inside Silencer.

4. Holding the Tool and its parts vertically, pass the Silencer Body down over the chambers. When all the way down screw on the Knurled End Cap. When the latter is tight, remove the Tool. Jar or shake the Silencer to settle things in place, and retighten the End Cap. The Silencer is then ready to use.

DIRECTIONS FOR DIS-ASSEMBLING SILENCERS

1. Insert the pointed or short end of the Tool into the threaded hole of Silencer and fit into center hole of first chamber. This makes you ready to push the Silencing Chambers out without injuring them. The Tool is essential for this and also for re-assembling.

2. Being sure the Tool is in place, unscrew Knurled End Cap from Silencer. When off, push Tool through. Silencing Chambers and Plain Washer should come out. If corrosion or rusting has occurred, and these parts stick, apply force, being careful to avoid injury to threaded end of Silencer Body. There are eighteen Silencing Chambers, and one Plain Washer. The latter is the first chamber in the Silencer, coming next the gun barrel.

INSTRUCTIONS FOR ATTACHING .22 CALIBRE MAXIM SILENCERS TO RIFLE BARRELS USING RIM FIRE AMMUNITION

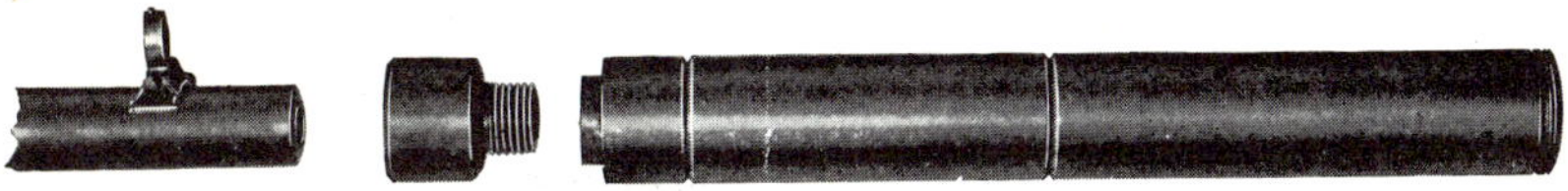

First: Examine end of rifle barrel and also interior of coupling, and make sure that there are no burrs nor chips nor dirt.

Second: Separate the shims which accompany the Silencer. These shims are intended to obtain just the right thickness of fit between the coupling and the rifle barrel. Insert in the coupling as many shims as are necessary to make the fit such that the coupling can be just started onto the end of the barrel about ⅛″ by hand. Take care that the shims do not get jammed or folded over. Unless the shims lie smoothly, the coupling will not be in perfect alignment with the barrel.

Third: When you are sure that you have a good tight fit, not so tight as to burst the coupling, nor so loose as to permit the coupling to be easily pulled off or twisted by hand, take a mallet and drive the coupling home on the gun barrel. You can tell when it is firmly seated home by the sound and the feeling. Your rifle is then ready to receive the silencer.

If the barrel is octagon instead of round, follow the same procedure as above, using a round coupling. The only point to be careful about is to use an extra shim so as to get an extra tight fit on octagon corners, and also be sure and rap the coupling all the way onto the gun barrel so that the end of the barrel bottoms in the hole in the coupling. This insures a gas-tight fit.

INSTRUCTIONS FOR THREADING RIFLE OR TARGET PISTOL BARRELS TO RECEIVE MAXIM SILENCER.

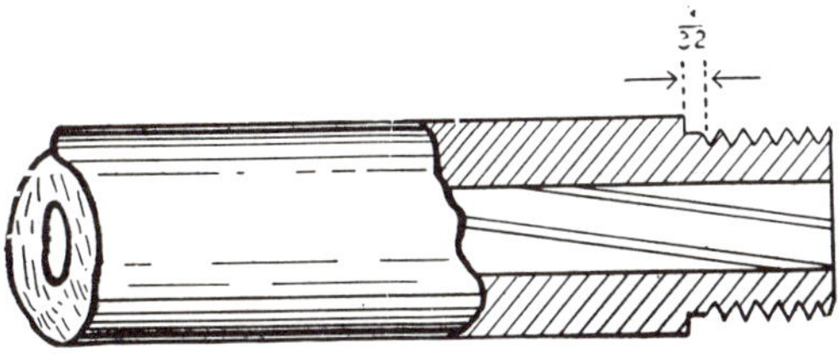

It is always much better to allow us to thread your barrel for you by means of the special threading tools which have been developed for this purpose. Any good tool maker or gunsmith can of course cut a thread on the end of a gun barrel, but without special tools it will take him more time and cost more money. It is very simple to pack your gun up and send to us or one of our dealers by parcel post. In a few days you have your gun back and it is properly threaded by an expert. For the benefit of those who desire to do this work themselves, we give the following information:

In threading any barrel in the lathe it must be mounted upon accurately centered plugs made to fit the chamber and the muzzle bore. This insures the running true by the bore. In non-take down guns the barrel must be taken out of the frame. (Our special tools avoid this.)

The threads for the various calibres are as follows:

1. .22 cal., both rim fire and high power........½ inch 20
   .25 cal. to .35 cal. inclusive, and corresponding millimeter sizes........$\frac{9}{16}$ inch 20
   .22 cal. 2 A coupling for small barrels........⅜ inch 24

2. The neck of the Silencer will be found to contain a counterbored portion ⅛ inch deep, and the outside diameter of the thread. THERE POSITIVELY MUST BE AN UNTHREADED SPACE ON THE BARREL TO FIT THIS COUNTERBORE. This is very important, as it is this fit that insures the correct alignment of the Silencer on the gun barrel. The illustration shows this unthreaded portion.

3. The fit of the threads should be just loose enough to enable the Silencer to be screwed on by hand without difficulty. The shoulder on the gun barrel should just bring the Silencer so that the groove in the bottom points straight down.

4. If the bore in the Silencer aligns with the bore of the gun barrel, the bullet cannot possibly touch the Silencer after it leaves the gun barrel. The bore of the Silencer is larger than the diameter of the bullet and ample clearance is therefore given.

DRIVE ON TYPE COUPLINGS are supplied with .22 cal. Silencers for rim fire rifles and target pistols.

THREADED TYPE COUPLINGS are necessary to attach the Silencer to very small diameter .22 cal. rifle barrels and target pistols. This coupling is known as a 2A, and requires that the end of the barrel be threaded with a standard ⅜" x 24 thread. When this coupling is screwed on the barrel tight, the Silencer may be attached or detached at will.

In the case of the Marlin Model 39 .22 cal. rifle where the magazine tube is very close to the barrel, leaving no room for the coupling, it is necessary to cut a standard ½" x 20 thread.

No couplings are supplied with Silencers for center fire rifles, which are always high power. Barrels must be threaded.

LIST OF .22 CALIBRE MAXIM SILENCER COUPLINGS

In the table below will be found a list of the different makes of .22 cal. rifles which can be fitted with a Maxim Silencer by the type of Coupling indicated in the margin preceding the name of the rifle.

Coupling	RIFLE	Model	Barrel
A	Winchester Single Shot	1902	Round
A	Winchester Automatic	1903	Round
A	Winchester Thm. Tgr. s.s	1899	Round
A-Flat	Winchester Repeater	1906	Round
B	Winchester W.R.F.Rptg.	1890	Octagon
B	Winchester S.S.bolt act'n	1904	Round
E	Winchester Single Shot	No. 1	Octagon
G	Winchester S.S. Musket	1887	Round
G	Winchester Bolt Action	52	Round
A	Remington Repeating	12A	Round
A	Remington Automatic	16	Round
B	Remington Single Shot	4	Octagon
B	Remington "Gallery" Rpg	12B	Octagon
B	Remington "Target" Rpg	12C	Octagon
2B	Remington Single Shot	6	Round
F	Remington Single Shot	R. W.	Octagon
B	Remington N.R.F.	1919	Octagon
A	Remington Automatic	24	Round
A-Flat	Marlin Repeater	1892	Round
2B-Flat	Marlin Repeater	1892	Octagon
A-Flat	Marlin Take-Down Rpr.	1897	Round
2B	Marlin Take-Dn.Hf. Mg.	1897	Octagon
2B-Flat	Marlin Take-Dn.Full Mg	1897	Octagon
2B	Marlin Repeater	32	Octagon
B	Marlin Repeater Hf. Mg.	20	Octagon
B-Flat	Marlin Repeater Full Mg	20	Octagon
K	Marlin Repeater	29	Round
2B	Marlin Repeater	38	Octagon
2BF	Marlin Repeater	37	Octagon
A	Stevens Little Krag	....	Round
A	Stevens Crackshot	16	Round
A	Stevens Little Scout	14½	Round
2B	Stevens Ideal	044½	Round

Coupling	RIFLE	Model	Barrel
2B	Stevens Ideal	56	Round
2B	Stevens Marksman	12	Round
2B	Stevens Ideal	40	Round
2B	Stevens Vernier	40½	Round
2B	Stevens Off-Hand Pistol	35	Round
B	Stevens Favorite	27	Octagon
E	Stevens Ideal	44	Octagon
E	Stevens Ideal	44½	Octagon
E	Stevens Armory	414	Round
E	Stevens Semi-Military	404	Round
E	Stevens Ideal Range	45	Round
F	Stevens Gallery	80	Round
K	Stevens Favorite 1915	17	Round
F-Flat	Stevens Visible Repeater	70	Round
A	Savage	1909	Round
A	Savage, Automatic	1912	Round
B	Savage, Repeater	1903	Octagon
B	Savage Bolt Action	1905	Round
B	Savage Repeater	1914	Octagon
2B	Savage	1904	Round
2B	Savage Repeater N.R.A.	1919	Round
2B	Savage	1911	Round
F	Savage Sporter	1922	Round
F	Savage Sporter	23A	Round
A	Hopkins & Allen	722	Round
2B	Hopkins & Allen	822	Round
A	Meriden	J-10	Round
B	Meriden	K-15	Octagon
F	Springfield	1922	Round
B	Browning Automatic	....	Round

The following guns take the 2A Coupling. Barrel must be threaded.

Geco .22 cal. rifle.
Smith & Wesson .22 cal. target pistol.
Reising .22 cal. automatic target pistol.
Colt .22 cal. automatic target pistol.
Stevens .22 cal. Reliable Pistol.
Stevens .22 cal. Diamond Pistol.

No couplings are furnished for attaching Silencer to the following guns. Barrels must be threaded.

Stevens .22 cal. No. 10 pistol.
Marlin .22 cal. Model 39 rifle.

NOTES

1. The action of powder gases is to corrode. It is well to reduce this action as far as possible by lightly coating the interior of Silencer and the interior parts with graphite grease. This can be obtained at any automobile accessory store. This will make it easier to take Silencer apart if it has been allowed to stand a long time after using.

2. If interior parts are allowed to corrode and it is difficult to push them out, the only thing to do is to force them by hammering the end of the Tool, or by using a vise. Care should be taken that the threaded end of the Silencer tube is not bruised. If chambers are injured new ones can be procured.

3. It is very much better to use the Tool provided for taking apart and assembling. This only costs fifty cents and it will save its cost many times over, as chambers are almost sure to be injured unless Tool is used.

4. Sometimes a Silencer will be put on the gun crooked. By this, we mean that the hole in the Silencer will be out of line with the hole in the gun barrel. This obstructs the clear flight of the bullet. It causes inaccurate shooting or injury to the Silencer. When a Silencer is first attached to a gun it is wise to test its alignment by means of a straight rod which may be thrust down through the Silencer and the gun barrel. If this rod is straight and if it is a fairly snug fit in the gun barrel, it will indicate at once whether the hole in the Silencer is in line with the hole in the gun barrel. The rod should pass through the Silencer without touching. If it does touch, the Coupling should be taken off and put on properly, or some other correction made. The gun will not shoot properly unless the bullet has a clear flight through the Silencer without touching.

5. Always keep the Knurled End Cap on outer end of Silencer screwed up tight. Also see that the Silencer is screwed onto the Coupling tight.

6. If your rifle is a Winchester or Marlin with tubular magazine, all you have to do when you want to load the magazine, is to slightly spring the magazine tube which will allow it to be withdrawn.

7. Never leave your rifle standing around with the Silencer on. It discolors the bore of the rifle under certain circumstances. It is much better to remove the Silencer after using the rifle. Take care to thoroughly clean your rifle bore after shooting, and then give a few strokes with a well oiled rag. This insures keeping the bore bright and clean, and also preserves the accurate shooting. If these instructions are followed, the rifle barrel can be kept accurate for many years.

8. If a rifle shoots inaccurately when the Silencer is on, it is a sure sign that the bullet touches the Silencer when passing through. It should not do this. The hole in the Silencer is much larger than the bullet. If the bullet touches, then it is a sure sign that the Silencer is not on straight. The cause of the Silencer not being on straight is usually the fact that the coupling was not put on carefully. The Silencers are accurately bored with the greatest precision, when they are in place upon what corresponds to a gun barrel. The way to correct bad alignment is to drive the coupling off with a piece of hard wood, and re-attach it properly. If a rifle with a Silencer shoots high, it shows that the bullet is touching the bottom of the Silencer. If the shooting is low, then the bullet touches the top of the Silencer bore. If to the right, it is touching on the left, and to the left, it is touching on the right. Usually the bullet just barely ticks, when the shooting is inaccurate, and if care is taken the trouble can be corrected by taking a small round file and filing a little at a time, until the shooting is correct. Only a little should be filed, and only at the place where the bullet is touching.

9. There is a lot of difference in ammunition. For example, .22 SHORT SMOKELESS gives good results both as to silence and cleanliness, but of course does not hit hard nor give long range. .22 LONG RIFLE SMOKELESS gives the best results of all. It is quiet, hits hard and is accurate at long range. It is distinctly the best of all ammunitions for use with Silencer. .22 LONG is about the worst, being noisy and inaccurate. It ought not to be used with the Silencer. .22 LESMOK makes more noise than smokeless and is also more dirty. It can be used but it is necessary to wash the Silencer out with hot water after using on account of the large amount of solid matter in the burned gases. In smokeless powder, there is practically no solid matter, and the Silencer consequently keeps clean for a long time.

10. When the Silencer is used on automatic rifles, extra care must be taken in the case of ammunition. The automatic breech opens so quickly that unburned powder grains will blow back into the action of the rifle if certain makes of ammunition are used. In other makes where all the powder burns, there is no trouble, and a very fine action results. If you have trouble with unburned powder in the action of your automatic rifle, try another brand of ammunition.

11. Round couplings are used on octagon barrels. They can be fitted very successfully if extra pains are taken in driving the coupling onto the barrel. An extra tight fit is necessary, and enough shims must be used to insure a very tight fit. In order to avoid leakage, it is also necessary to be sure that the coupling is driven all the way home, and bottoms against the end of the gun barrel.

12. We do not manufacture Silencers for shotguns larger than .45 calibre.

13. Do not blame the Silencer for not controlling ALL NOISES. All the Silencer can control is the report noise. It cannot silence bullet flight noise, nor the noise made by the impact of a bullet against a tree trunk or back stop, nor the mechanical noise made by the breech mechanism of an automatic rifle or target pistol. Frequently the Silencer is blamed for not controlling these noises, so that we have to mention it.

14. The .22 cal. W. R. F. has caused trouble because many people expect it to shoot as quietly as .22 long rifle. It will not because the bullet velocity is above the critical and the .22 long rifle is below the critical, and consequently the former makes bullet flight noise, whereas the latter does not. The Silencer annuls the report of each.

15. On high power rifles the Silencer alters the targeting on account of the added weight changing the vibration characteristics of the barrel. This can be provided for by adjusting the front or rear sight.

16. It is positively necessary to keep the Silencer screwed up tightly on the gun barrel. Unless this is done the Silencer will get out of line and the bullet may touch when passing through and injure the silencing chambers.

Every Maxim Silencer is sold with the understanding that it will practically annul the report noise and also reduce the recoil over sixty per cent. but that it will not silence the mechanical noises made by the breech mechanism, nor the noise made by the bullet in its flight through the air, nor the noise made by the impact of a bullet against a back stop.

HOW TO FIND IF YOUR SILENCER IS SILENCING

In a single shot or repeating rifle there are two separate noises made when shooting a high power rifle.

1. One is the noise made by the "REPORT" or that noise caused by the sudden liberation of the powder gases into the air.

2. The other noise is that made by the BULLET'S FLIGHT.

The "Report" Noise is practically entirely annulled by the Maxim Silencer. The Bullet Flight Noise is not affected by the Maxim Silencer, because it is produced out beyond the gun.

The noise which indicates the general location of a gun is the "Report" Noise. The Bullet Flight Noise does not give any indication of location, whether North, East, South or West.

It is, of course, true that a gun makes the same noise no matter in what direction it is pointed.

It is also a fact, although not so well known, that the bullet makes a sharp crack in its flight through the air if its velocity exceeds a certain figure. This noise is called "Bullet Flight Noise." It occurs out in the air beyond the gun. It can only reach the ears of the shooter by reflecting back from some object. The ground, trees, or buildings are such objects.

One way to demonstrate the bullet flight noise, and incidentally to prove whether the Silencer is working right, is to fire into a bag of soft earth or sand, holding the muzzle of the gun with Silencer within three or four inches of the bag. There is small opportunity for bullet flight noise to develop across this small distance, and the noise heard is principally that made by the impact of the bullet against the sand.

If the Silencer is working right, this noise should be a heavy "grunt." After trying this, step back twenty-five feet and fire into the bag of sand. A sharp "crack" will be heard. This is BULLET FLIGHT NOISE. The gun report is the same whether held up close to the bag or not. If you doubt it, try it.

Another way to demonstrate is to go out into an open field clear from all objects like trees or buildings. First shoot the bare gun straight up into the air, then also parallel with the ground. You will hear the report noise both times.

Now put the Silencer on and repeat. When you shoot up into the air, you will hear only a "soft" noise. This is what is left of the report noise. There is no bullet flight noise heard, because there is nothing up in the air to reflect the noise back to your ears. The noise goes on and on and never returns. Now, shoot parallel with the ground. You hear a "crack." If you shoot past separate trees or telegraph poles, you will hear the bullet as it passes each. Shooting down a railroad track, past a long row of telegraph poles, sounds like a clatter. Of course, the gun makes the same noise whether pointing up or down, so the noise heard in this last test is bullet flight noise.

Bullet flight noise begins when the bullet velocity exceeds the velocity of sound. It is approximately 1,075 feet per second. If you crowd any object through the air faster than this rate the air "breaks" something as the bow wave of a boat breaks when it is forced through the water faster than the water can get away from the bow. The tip of the whiplash "cracks" when you snap it vigorously enough to cause it to acquire a velocity in excess of 1,075 feet per second.

Any ammunition which has a bullet velocity below 1,075 feet per second, does not make bullet flight noise. Any ammunition above this figure does make bullet flight noise. It makes no difference how quiet the Silencer makes the gun, there will be a noise made when shooting, IF YOUR BULLET EXCEEDS 1,075 FEET PER SECOND VELOCITY.

.22 calibre smokeless Short or Long Rifle and .22 calibre Winchester Automatic have silent bullet flight. .22 Long and .22 W. R. F. have noisy flight. .25-20 calibre is noisy, and all high power cartridges with regular loads are the same.

Some ammunition can be had in modified loads. Should you, however, not be able to secure this, you can load your own shells to give the desired effect.

THE MILITARY ADVANTAGES of the Maxim Silencer are different than they are supposed to be by the average citizen. Instead of entirely eliminating all noise, the Maxim Silencer only eliminates the "REPORT NOISE." In addition to this, it also REDUCES RECOIL 75%, ANNULS FLASH when firing at night, and IMPROVES MARKSMANSHIP. These advantages are not to be proven by taking one gun and shooting a bullet into a nearby tree and listening to the noise made. They can only be demonstrated by using a body of soldiers and observing their improved efficiency under service conditions when they have the Maxim Silencer on their rifles.

EXPERIENCES WITH THE
MAXIM
SILENCER
By
Hiram
Percy Maxim

HIRAM PERCY MAXIM

Foreword:

The MAXIM SILENCER was developed to meet my personal desire to enjoy target practice without creating a disturbance. I have always loved to shoot, but I never thoroughly enjoyed it when I knew that the noise was annoying other people. It occurred to me one day that there was no need for the noise. Why not do away with it and SHOOT QUIETLY?

For nearly two years I sought for some way to check the powder gases from bursting into the air when the bullet left the barrel. This is what causes the objectionable report noise. I thought of every kind of a valve, vent, by-pass, expansion chamber, etc., and built some of them. None were satisfactory. It seemed quite a problem to find a hole through which the bullet could pass, but through which the powder gases could not pass.

One morning after my bath I noticed in the bath tub the miniature whirlpool that forms over the drain hole when the plug is pulled and the water starts to run out. There was the familiar little hole down in the center of the whirl and it started me thinking that here was an exactly similar case to my powder gas and bullet problem. Here was water in a bath tub, the drain plug pulled out, and yet the water was able to run out but SLOWLY because it was WHIRLING.

Why would not the powder gases act the same way as the water, if they were whirled? The whirling would give them centrifugal action precisely as it did the water, and cause a "hole" to form in the center just as the hole formed in the water. In a flash I saw the "hole" for which I had been looking for nearly two years.

I immediately made a little "whirling tube" which would catch the powder gases as they burst from the muzzle of one of my rifles and whirl it around vigorously. In the center I provided a hole for the bullet to pass through but considerably larger than the bullet so it could not touch The gases had no escape except through this central hole. Being central they could not possibly get out until they had SLOWED DOWN. This of course meant that they must come out gradually, and, in consequence, noiselessly. The first time I shot the rifle I was quite excited. I will never forget the sensation I experienced when I found IT WAS QUIET.

That was the birth of the Maxim Silencer.

The letters from sportsmen in all parts of the country, reproduced on the following pages, will give a more graphic idea of the uses and advantages of the Maxim Silencer than I could in my own words. You will find every one of these letters exceedingly interesting.

HIRAM PERCY MAXIM.

My Own Experience in Teaching My Son to Shoot

BY HIRAM PERCY MAXIM

I had an interesting example of the advantage of the Silencer in target shooting in the case of my son Hamilton. He was nine years old at the time, and for a year had owned and shot remarkably well a little .22 cal. Savage rifle. One day some friends and I planned an afternoon on our 200-yard range with the U. S. Springfield service rifle. Hearing of it, the boy asked to be allowed to go along. Always willing to encourage his interest in the manly sports, I consented.

After watching the shooting for a while, he suggested that he would like to try it. At first I would not consider it, fearing the child would be hurt. He was only nine years of age and I feared to let him shoot such a powerful gun even with the Silencer. The more I thought of it, however, the less dangerous it seemed, and finally I decided to make the experiment.

The rifle was too heavy for him to hold and he had to shoot kneeling. I gave him one cartridge to begin with and after warning him to hold tight and be careful I left him to work it out himself. He understood the sights. After much preliminary squirming he finally found a position which suited him, and sighting as carefully as an expert, he pulled. I half expected to see him burst out crying. He did nothing of the kind, however, but quietly looked up and waited for the shot to be marked. Judge my astonishment when it was a 4.

Believing firmly it was an accident, I asked him how he felt and when he answered "all right," I gave him another cartridge. He repeated the operation, only this time it was a 5—a bull's eye. Wondering what it meant, I gave him a clip of five cartridges and stood back to watch. He got two more 5's, a 4, one miss and a 3. To complete his ten, I gave him three more cartridges with which he got a 4 and a 3, giving him 33 out of a possible 50.

Needless to say, the young man was heartily congratulated by all present. His target was dated, signed by himself in his own scrawling handwriting and witnessed, and afterwards photographed. Twenty years hence, when he grows up, it will be an interesting relic.

Afterward he shot several more strings with us, but being flustered by the attention he had received, did not do as well. He must have shot thirty or forty rounds altogether. When we started for home, he was asked how his shoulder felt. He said his shoulder was all right, and appeared surprised at the question. Little did he suspect what he had been saved. Without the Silencer, he would have been seriously bruised and his tender young ears would have rung for hours.

Ten Cent Piece Covers 10 Shots

"The .22 calibre Maxim Silencer I bought some time ago for my Stevens Pope Rifle, has given entire satisfaction.

I have made 10 consecutive shots at 50 yards that could be covered by a ten-cent piece.

I would recommend this Silencer for indoor use, to all marksmen."

E. S. M., Waynesboro, Pa.

Showing the Boy

Noiseless
Indoor Shooting

His First Crow Shoot

"I certainly had a funny go with one of your Silencers last Sunday. J—— gave me a Winchester automatic .22 cal. for Christmas, with a Silencer attached, and it sure is a beauty. This makes two Silencer outfits in the family now. T., my fat brother-in-law, was with us over Sunday, and I thought it was about the proper thing to show him what a lot of fun can be had shooting crows with a Silencer. In the winter there is nothing else around here to shoot, and I have worked it out so that it is a pretty live sport.

Well, T—— said he was willing, so we got into my car and ran out to the place I have, which is on the line the crows take every afternoon on their way to the rookery. Every afternoon about four o'clock they begin and there is a steady flight of them for an hour or more. There must be thousands and thousands. I have a little blind fixed up in a pasture where there are a lot of stunted cedars, and it makes a fine stand.

We got in the blind about four and I gave T—— the repeater while I took the new automatic. It was the first time I had taken the latter out. Somehow or other the crows did not appear as soon as usual, or else we were a little early, for T—— commenced to jolly me about bringing on my crows. He was all ready all right, and if I would attend to my part of the contract, the ball could open.

Finally I sighted the advance guard. I got the old crow call going, much to T——'s amusement, and in about a minute I saw it had been heard. The whole line veered and started toward us. T—— didn't catch on until they were right overhead, circling and cawing like mad. Then you should have seen him. He is not much of a sport, and probably has been hunting about twice in all his life. He is fat, and likes a hammock, a good book and a black cigar more than tramping in the open. But he has all the good old instincts just the same, although they are dormant most of the time.

Noiseless Outdoor Sport

He had my repeater which carries some 12 or 13 cartridges, and I had told him how to load and work it. He had a little box of cartridges opened on the ground beside him, and as it was pretty cold, he had his gloves on. When T—— caught sight of the cawing birds circling overhead, hundreds of them all within easy range, he jerked his gun to his shoulder with a snap, spat out his cigar butt, and went into action with a vengeance. He astonished me. All the fighting blood of his noble ancestors rose up in him and he went at it as though his life hung on the results. He pumped in fresh shells and blazed away and squirmed around, trying to kill the whole flock at once until he nearly upset the blind. I got to laughing so I simply had to give up.

Finally he came to the last cartridge and discovered that his magazine was empty. Did he stop? Well, I guess not. He dove into the box of cartridges and in a frenzy began to reload. Picking them up one at a time wasn't fast enough, so he just poured them out. Then his gloves got mixed up and he spilt a lot out on the ground.

Then off came the gloves and such a scramble you never beheld in all your life. I had stopped shooting to watch him and the birds stopped coming, or saw us, so by the time T—— was ready again there was nothing to shoot.

He was heartbroken. He just thirsted for blood. Turning to me he said, half mad, "Why didn't you shoot?" I told him I had done my best, having got three out of ten shots, which wasn't so bad. "Come off," said he, looking me in the eye for a joke. When I pointed out the dead crows out in the field you should have seen good old T——'s face. *He had not heard my gun, notwithstanding he was in the same blind with me.*

Well, we had several more "battles" and each time T—— almost went mad. He shot away every blooming shell he had and never hit a bird. All he seemed to want to do was to wade in and shoot as fast as he could pump the gun. I got a little better results with the automatic, probably because I was less disturbed by not having to work the "pump" of the repeater. And it takes some shooting, let me tell you, to get a crow on the wing, even though he is circling overhead. For me it beats trap shooting all hollow.

But just think of it—crow shooting Sunday afternoon with no noise, in fact, not enough for the man in the same blind to hear your gun. Without the Silencer this would be absolutely impossible. One shot, and good-bye crows."

W. T. S.

Improved His Marksmanship 25%

"Since writing you I have purchased a .22 cal. Maxim Silencer of Jos. Weigand, a sporting goods dealer of Merrill, Wis. This Silencer on my target pistol completely eliminates the noise of shooting and there is no more noise than is made by the action of the gun. It has also improved my shooting about 25 per cent., as it counteracts the tendency of the gun to jump or fly upward at each shot. I would not take $50.00 for it, if I could not get another.

I have not tried a Silencer on a high power rifle yet as I disposed of the rifle I had, but I will purchase another before next hunting season and you may be sure it will be equipped with a Silencer."

R. R. J., Irma, Wis.

Splendid for Noisy Cats

"I want to tell you how your Silencer helps us keep house. We live in a big apartment house and there are neighboring back yards which are regular battle fields for cats. These half wild creatures nightly get into political arguments which disturb us so much we lose sleep.

My wife, who is a nervous invalid, but nevertheless gifted with a fine working imagination, suggested to me that I buy a Maxim Silencer and shoot the troublesome animals. Being something of an outdoor man and fond of hunting, the idea struck me as pretty good. I bought a Silencer for my little Remington, and put it on easily—thanks to the little coupling which came with it.

The first night after getting the thing fixed up I had a good chance to use it. We had retired and were just getting to sleep when a couple of cats started up. I immediately got up, and finally located them right down on the grass of a yard. Knowing the little .22 short cartridge would be perfectly safe fired into the grass and earth from above, I got out the little rifle and with considerable excitement got a bead on the biggest one of the cats. I pulled the trigger, there was a little click and the old cat went end over end a couple of times and then lay still. The other cat paused a moment and then withdrew in haste, and we were left in quiet.

I suppose the people who found the dead cat put it in the ash barrel for me. It is great having your game retrieved this way. Since then I have done away with just ten cats who have disturbed our sleep at night. Instead of being in trouble all the time over the noise, we now have quiet nights or the means to make them quiet if they get noisy.

Anyone understanding how to shoot safely, and living in a neighborhood where cats are troublesome, ought to get a little .22 cal. rifle and a Maxim Silencer without delay."

A. L. M.

Thought It Had Missed Fire

"You will be interested in a little incident I had up in Canada on a recent fishing trip.

I took along a .22 cal. rifle which I fitted with one of your Maxim Silencers. I always take this little outfit with me everywhere I go, just for the pleasure that can be extracted from shooting at a mark. One morning before breakfast at the camp, I stepped outside with one of the other members of the party to take a few shots at whatever there was around. Frequently I get a little something which goes well on the table. I had turned the gun over to my friend to try, when a large hawk appeared overhead, circling and looking for his breakfast.

Pointing to it as a fine shot and a good thing to get rid of, he up with the gun and begins shooting. The first shot he took he landed, as I could distinctly see. Evidently he was not aware that he had hit the hawk, possibly because of his astonishment at shooting a noiseless gun for the first time. He brought the gun down from his shoulder and looking at it, he said, "missed fire." In the meantime the hawk's body had fallen on the far side of a tree. "Missed fire?" I replied, "how about that?" pointing to the hawk's body on the ground.

Needless to say he was amazed. He had given no thought to the hawk, thinking the gun had missed fire. He had to shoot several times before he could believe his own eyes, that he was actually shooting a real .22 long rifle cartridge. I thought you might be interested in this little experience."

J. E.

He Got His Deer

"I got a deer this season which I would not have got if it had not been for your Maxim Silencer. It happened like this:

I came out on to a little pond, and across on the other side I saw a deer drinking. I had my .30-30 and immediately brought it up and let go. I missed, holding too high. The deer did a strange thing.

It jumped, hearing the bullet crack overhead, but did not seem to know which way to run- to avoid the noise. As you know, bullet crack always sounds right over your head, no matter which way the bullet is traveling. There is no idea as to where the gun is.

This deer was confused. He finally backed into bushes before I could get another good chance at him. I waited to see what would happen, and in a minute or so out he came again. He looked around and came all the way out, evidently not entirely scared, but just curious. I let him have it again, and strange to say, I missed the second time. I held on his shoulder and he moved just as I pulled. This time he jumped again, but did not move away, seemingly trying to study out that little "cracking" noise.

I got another shell in and this time I got down to business and landed him. The distance was not more than 100 yards, and you can imagine how much chance I would have had at that deer without the Silencer. The first loud bang of the report would have sent him scurrying, never to return.

I had several other experiences with the Silencer, which I think would interest you. I am convinced it is a very good thing from the humane standpoint and also for saving the game, although many people who have had no experience with it think otherwise. With a Silencer a hunter gets his first animal, usually. He has more chance to. This means there are less wounded animals which get away and die. We all know how often we find a poor creature's body with all the signs of having crawled away wounded and left to die slowly. The loud report noise of the regular gun prevents a second or a third shot, which in many cases are necessary to kill."

W. G.

Muskrat Hunting

"Your Silencer certainly is a great thing for muskrat hunting. I had an experience the other day which I am going to write you about.

At this season our river comes up and floods the banks and forces the rats out of their holes and we get them with small rifles and shotguns. A friend and I were out after them in a boat and I had my .22 rifle with one of your Silencers on it so as to keep down the noise. I saw a rat on the bank and after getting the boat as near as we dared, I fired. I failed to connect, probably because the boat was not steady. The bullet plunked into the mud back of him. Of course I expected that would be the last seen of the rat.

But this was where the Silencer came in. Instead of diving, that rat, hearing only the soft plunk of the bullet in the mud back of him, turned to look at what was behind him, thinking that was where his danger was located. I caught on right away and getting in another shell I took more careful aim and got him. I would have lost him without the Silencer.

Some time I will write about a queer experience I had shooting skunks with the Silencer."

Sparrow Shooting

"Please send me catalog showing present method of attachment of Silencer. I want one for a .28-30 Stevens barrel which is not standard either in muzzle diameter or taper. The Silencer which I have on a .44 cal. shot barrel is about the most satisfactory implement I ever possessed.

The total bag of English sparrows on my premises of about five acres, between two streets, since January 1st, is less than a dozen while on the premses of an adjacent neighbor where they have not previously been shot, I have killed about a hundred and fifty in the last month.

The consequent increase in song birds and other insect eaters is notable, and is plainly shown in condition of fruit trees and shrubbery. This shooting would be entirely out of the question in this vicinity without the Silencer.

It is especially to be noted that this muffled shooting does not in the least degree alarm the other birds." F. W. W., Norwichtown, Conn.

Miniature Trap Shooting

"I am sending you a couple of photos of our Noiseless Trap Shooting Club which I trust will interest you. We use .22 and .32 shot cartridges. in smooth bore rifles fitted with your wonderful Maxim Silencers. We also use regular Blue Rock Clay "birds" and a regular Leggett trap. The latter is mounted upon a bracket so that the "bird" is thrown nearly straight up. We stand back twenty feet from the trap.

This outfit enables us to enjoy trap shooting on the front lawn without disturbance or danger. The Silencers stop the noise and the shot cartridge is perfectly harmless beyond 200 feet. Some of us have become crack shots, and we think it great sport. Incidentally it is much less expensive than regular trap shooting, and it is always convenient. We can set up the trap any time in almost any place. We propose trying .32 single shot smooth bore pistols next, with Silencers. If it works out all right we will have pistol trap shooting, something entirely new."

Noiseless Trap Shooting

A Slender Young Lady's Experience

"I have just got back and today have been trying out that Featherweight Savage .303 rifle which you fitted with a Silencer, and must say that I am very pleased with the result. The gun makes a bit of row with the high steam cartridge, but it does what I particularly wanted: reduces the recoil. And with those cartridges which you supplied, it does away with both recoil and noise, much better than when using the auxiliary cartridge and revolver ammunition. It looks almost ridiculous to pull the trigger and see a squirrel drop with less noise than snapping a primer in an empty shot gun shell.

I am sorry that I did not have a rifle fitted with a Silencer long ago. It is the first one in this district and appears to be attracting quite a little interest among the local shots, quite a few having been patiently waiting for me to come and try it out.

I ordered a .22 cal. rifle to be fitted with a Silencer, on my way up, which you may have heard of by now. I will be pleased to answer any one inquiring about a Silencer around here or let them see it used, should they care to come along.

I may say that the rifle will be shot mostly by a slightly built lady, who is enthusiastic over the idea of being able to kill big game without having her shoulder battered to the various colors of the rainbow by her high power arm."

J. B., Flagstone, B. C.

Bringing Ducks Up to the Gun

"Here is an experience with one of your Maxim Silencers which you will probably find hard to believe, notwithstanding it is an actual fact.

I was duck shooting from a boat which I had pushed in among the reeds out of sight. A flock of ducks came in and settled in the water considerably beyond the reach of my shotgun. I had figured that one of your Silencers on a .22 cal. rifle would be a good thing to have around when shooting ducks. I decided to try it this time and see if I could not bring those ducks nearer in.

I picked up the .22 and shot out beyond the ducks so that the bullet would hit the water on the far side of them and far enough off to wake them up but not frighten them into the air. They immediately took notice when the bullet splashed in the water and began swimming away from the disturbance, and toward me.

After a while I sent another bullet out and kept repeating the operation every minute or so. I found that I could steer these ducks perfectly. They could not hear the gun and of course I could bring them right up toward me. I really brought them up within easy range of the shotgun and got one with each barrel.

It seems like a fairy tale to write, but I vouch for the absolute accuracy of the above."

R. E. P.

Killing Crippled Ducks

"I am one of those who enjoy duck shooting and one of my difficulties has been killing crippled ducks. There are always several of these in a day's shooting which are not killed, but seriously crippled. They fall into the water and if they are in range of the shotgun you have to blaze away until you kill them. Sometimes you do it and sometimes you don't. Anyway, you kick up an awful row, scare new birds from coming in, and waste a lot of ammunition.

I have found that a .22 cal. rifle with a Maxim Silencer just fills the bill for killing these cripples. It allows you to get them when they are out beyond the range of the shotgun, and there is no noise to frighten away other flocks. The best cartridge to use I find to be the .22 smokeless Long Rifle. I can get a duck practically every time at 100 to 125 yards. You ought to tell duck shooters about this. Most people dislike to cripple the poor creatures and let them get away to die a lingering death."

A. B. T.

Training Recruits

The following is taken from a letter received from an Army Officer:

"Enclosed you will find the record of five men in my company who have given us a lot of trouble. Two of them are recruits and three are old men. All five are what we call "chronic flinchers." They have been the despair of their officers and comrades, being so gun shy as to be incapable of making a respectable score on the range.

Believing that Silencers would help them, I secured five and fitted them to their rifles. The scores tell the rest of the story.

You will see that the first string beat the best previous records ever made by any of these five men. You will also see that after finishing the programme with the rifles fitted with Silencers, these men were able to also beat the previous averages with regular rifle. I ascribe this to learning the knack of shooting without flinching. This knack was a sealed book to these men before they used the Silencer.

There is no doubt but what much can be gained by starting recruits in with rifles fitted with Silencers and thus teaching them the principles of proper holding and trigger pull. The greatly reduced recoil, and the annulment of report effected by the Silencer, prevent forming the flinching habit. I consider these results as very important."

The Silencer works perfectly on the .22 Winchester automatic rifle.

The Silencer is not recommended on high power auto-loading rifles nor shotguns.

Shot loaded cartridges up to .45 cal. work perfectly in the Silencer.

The Maxim Silencer will not silence a revolver nor an automatic pistol.

First Use of the Maxim Silencer in Actual War

This report was made to me by the personal representative of a Mexican officer who for the past two years has figured in active hostilities. The report gives the first authentic information regarding the use of Maxim Silencers in actual battle.

It seems that a number of Maxim Silencers were purchased in one of the border cities on the American side. They were evidently used extensively in the first revolution on Madero's side, and it seems that they are again in use at the present time. When I asked about their military value, I was informed that my Silencer's chief value was altogether different than I had predicted.

Instead of its chief value being the concealment of position, I was told that it was the favorable effect upon the soldier. The latter, when furnished with a rifle equipped with Silencer, became possessed of the idea that he could take greater chances without injury to himself, and always shot more carefully. This, according to my informant, was a very valuable thing indeed. He said that it tended toward making every man a sharpshooter. I was asked if I had expected this, and I was forced to acknowledge that I had not thought of it, although I could easily see how it could be so.

I imagine this will surprise several military experts, as it surprised me. No one to my knowledge predicted this.

Another curious thing was the reported effect of the Silencer upon the nervous and physical organization of the soldier. It seems that in consequence of the terrific report and recoil of many of the rifles used, that the men acquire a state of excitement, which after several hours produces serious fatigue. In this condition the average man cannot be counted upon. It was stated that the men with the Silencers could always be counted upon at critical moments, even late in the day. The reason given was that these men had not been subjected to terrifying noise and continual recoil pounding.

This condition was predicted by some of our United States officers after the first Maxim Silencer tests.

Another thing that came out as predicted according to the report, was the effect upon marksmanship. I asked about this particularly, and the reply was that the Silencer gives the soldier the idea that he is safer and he can be more careful in aiming, and as a natural result his marksmanship is better. It was claimed that with the Silencer the soldier takes an interest in hitting what he aims at. If this is so throughout the rank and file of an army, it is a matter of the greatest importance.

The prediction made by most military experts regarding improved control of the firing by officers was borne out. The fighting in Mexico seems to be generally open order. The men are scattered and not in solid masses.

It was said that officers have great difficulty in controlling the fire of these men because of the noise drowning out the sound of their voices. When the men having Silencers were grouped together, their control was said to be perfect. Officers' orders could always be heard no matter how rapidly they were shooting. It is interesting to note that this exactly bears out the predictions made by the U. S. School of Musketry. Confusion on the part of the enemy regarding location of firing line and the number of guns against them, and also the reducing waste of ammunition, were mentioned. The report was of great interest as it constitutes the first word from the actual battlefield.

HIRAM PERCY MAXIM.

The Deadliest Firearm Ever Invented

Your Question is Probably Answered Here

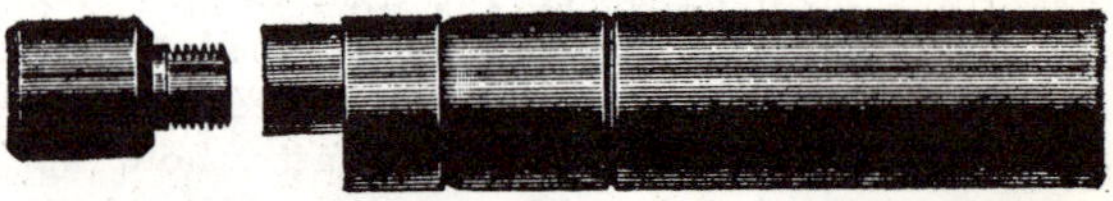

.22 Calibre Rifles

This is an accurate illustration of the .22 calibre Silencer and its coupling.

The coupling simply has to be driven tight onto the end of your gun barrel. Once on, the Silencer can be attached or detached at will by a single twist of the wrist. A small sleeve or cap is furnished which can be used to cover the threads on the end of the coupling when the Silencer is off, thereby giving the gun its usual appearance.

Different-sized couplings are made for different-sized gun barrels. When ordering state MAKE and MODEL of your rifle, and whether barrel is ROUND or OCTAGON, and the proper size coupling will be furnished.

Model 1912, 1 in. x 4¼ in., wt. 5 oz., $5.00 with coupling.

This Silencer will make your gun noiseless. It will double the pleasure of your shooting and your advantage in hunting and target practice. Once you have enjoyed the experience of SHOOTING QUIETLY, you will never be satisfied with the old-fashioned noisy report.

.25-20 and .32-20 Rifles

As with the .22 cal., these Silencers are attached by means of couplings which simply have to be driven tight onto the end of the gun barrel. Once the coupling is on, the Silencer can always be attached or detached by a single twist of the wrist. Different couplings are made to suit the various sizes of gun barrels. When ordering, state MAKE and MODEL of your rifle, and whether barrel is ROUND or OCTAGON, and the proper size coupling will be furnished.

The dimensions and weights are as follows:—

.25-20, 1 in. x 5 in., wt. 7 oz., $7.00 with any size coupling.

.32-20, 1⅜ in x 6¼ in., wt. 12 oz., $7.00 with any size coupling.

Either one of these Silencers will absolutely annul the noise made at the muzzle of the gun by the report and also take out practically all of the recoil. With ordinary ammunition, this does not give strictly noiseless shooting because the bullet from the usual cartridge has a velocity high enough to make a noise of its own in its flight through the air. By sacrificing some of the velocity, this bullet flight noise may be avoided, and strictly noiseless shooting obtained. We furnish information regarding loading it yourself.

.22 Savage High Power "Imp"

This silencer is attached to the gun barrel by means of a coupling which slips onto the end of the barrel and locks back of the front sight by means of a clamp band. The Silencer can be attached or detached at will by a single twist of the wrist. A small cap covers the threads on the end of the coupling and gives a neat appearance to the barrel when the Silencer is off.

The size and weight are as follows:—

Diameter 1 in., length 5¾ in., weight 7 oz. Price of Silencer only $6.00. Price of Coupling only $2.50.

This Silencer is especially good on the Savage "Imp." The gun has an extremely sharp report and the Silencer entirely does away with this. It also takes out the last bit of recoil. It doubles the pleasure of shooting the gun. It, of course, does not stop the noise made by the bullet in its flight through the air. As with other high power rifles, this bullet flight noise is always heard with regular ammunition, notwithstanding the fact that the gun report is entirely eliminated by the Silencer. By sacrificing some of the bullet velocity, strictly noiseless shooting can be obtained.

.25-35 Rifles

For Winchester, Savage, Stevens and Marlin rifles of this calibre, a Silencer and Coupling is furnished. The coupling drives tight on the end of the gun barrel after removing the front sight from the latter. A new front sight is furnished in the Coupling to take the place of the one driven out of the barrel.

The Silencer can be attached or detached at will by a single twist of the wrist. A small cap covers the threads on the end of the coupling when the Silencer is off.

The size and weight are as follows:—

Diameter 1 in., length 5¾ in., weight 8 oz. Price of Silencer only $7.00. Price of Coupling only $2.50.

The Silencer entirely stops the noise made at the muzzle of the gun by the report and also takes out 75 per cent. of the recoil. With ordinary ammunition, strictly noiseless shooting is not obtained, because the bullet from the regular cartridge has a velocity high enough to make a noise of its own in its flight through the air. By sacrificing some of the velocity, this bullet flight noise can be avoided and strictly noiseless shooting obtained. We furnish information regarding loading it.

The Silencer improves marksmanship and the advantage in hunting or target practice to an extent that is difficult to believe until it has been tried. With the reduced velocity ammunition, target practice can be enjoyed anywhere without creating the least disturbance or attracting attention.

.30 Calibre Up to .45 Calibre Rifles

For Winchester, Marlin, Savage and Stevens and all other makes of repeating or single shot rifles in any calibre between .30 and .45. Silencer and Coupling is furnished with either SIGHT TYPE "DRIVE-ON" COUPLING or CLAMP TYPE COUPLING.

The "drive-on" type coupling is used on all rifles having front sight which can be removed from gun barrel, leaving a plain round barrel.

The clamp type coupling is used on all rifles having front sight mounted upon a base built into the gun barrel.

The Silencer attaches or detaches from the coupling by a single twist of the wrist. A cap is furnished which covers the threads on the end of the coupling when the Silencer is off, giving the gun barrel a finished appearance.

The size and weight are as follows:—

Diameter 1⅜ in., length 6¼ in., weight 12 oz. Price of Silencer only $7.00. Price of Coupling only $2.50.

On any one of the rifles mentioned above the Silencer absolutely annuls the noise made at the muzzle by the report—even in the most powerful, such as the Winchester cal. 405, or the Winchester Model '95 shooting the Government '06 cartridge. This is not only accomplished, but in addition the recoil is reduced 75 per cent. and changed from a kick to a softened push.

As in the case of the other high power rifles mentioned above, noiseless shooting is not obtained with the regular cartridge because the bullet of the latter has a high enough velocity to make a noise of its own in its flight through the air. This noise is, of course, not nearly as loud as the report noise. It can only be stopped by sacrificing enough of the velocity to bring the latter down to 1,100 feet per second, at which point the bullet flight noise always begins. We furnish information regarding loading.

The Silencer vastly increases the value of any of the high power rifles on account of the greater pleasure and interest in shooting and the greater use that can be made of the gun. Small game can be shot around camp without frightening away big game which may be in the neighborhood. Target practice can be enjoyed even at home, with reduced velocity ammunition, without the slightest disturbance being created. Marksmanship is improved because of the stopping of the report noise and reducing the recoil which stops the tendency to flinch almost unavoidable when shooting a high power rifle.

All couplings are round, whether intended for round or octagon barrels.

By threading the rifle barrel, any Silencer can be attached without coupling. We recommend this direct attachment wherever possible, as it saves both the weight and expense of the coupling.

One Silencer can be used on any number of different guns, provided the calibre of the gun is not greater than the calibre of the Silencer.

The Silencer does not injure accuracy, but actually improves it on account of the steadying action of the Silencer on the gun barrel. The bullet is not supposed to touch the Silencer in passing through and consequently it has the same clear flight whether the Silencer is on or off.

The best .22 cal. ammunition to use in the .22 cal. Silencer is regular smokeless Long Rifle. The cartridges furnished by some makers loaded with Lesmok powder have bullet velocities high enough to make bullet flight noise. The regular smokeless cartridge does not.

The Silencer can be cleaned by letting hot water run through it. A good way is to attach it to a hot water faucet and let the water run through slowly for two or three hours.

MADE IN U.S.A. BY

O. H. Brown
MECHANICAL ENGINEER

DAVENPORT, IOWA

FOREWORD

At the close of the world war the only firm that had been engaged in the manufacture of gun silencers had discontinued their manufacture and the only ones to be had were a few that happened to be in the hands of dealers scattered throughout the country and a quantity manufactured for the Government to be used on sniper rifles, which were sold to a few sporting goods dealers, one of whom engaged the writer to reconstruct them in order to make them adaptable to sporting rifles and pistols.

These silencers were advertised in various magazines and were readily sold. When this supply was exhausted, orders and inquiries continued to come in. This led me to believe that there was a good demand throughout the country, for a well made silencer selling at a moderate price. With this idea in mind I proceeded to design and make up a quantity in the small 22 caliber size and advertise them for sale. That this belief was justifiable was evidenced by the fact that the orders for the first few months exceeded my ability to turn them out.

These silencers are now manufactured in three sizes. The small size is intended for arms using the 22 caliber rim fire ammunition only. The second or medium size, is intended for such arms as those using the 22 W.R.F., 22 Hi-Speed, 25 Cal. rim fire, the 25-20 and 32-20 center fire ammunition. The large size is for calibers from 22 Hi- power to 38 inclusive. The illustration at the top of this page is one-half size cut of the Hi-power silencer.

The purpose of this folder is to give a description of these silencers, the prices of each, and the various arms on which they can be used. It is also hoped the paragraphs on the following pages will give any other information that the prospective buyer would wish to know.

O. H. BROWN

N. B. This folder superceeds all others and all sales are governed by the prices and conditions named herein, effective June 1st, 1931.

IS THE SILENCER PRACTICABLE?

The silencer is a practical device for use on repeating and automatic rifles and pistols, and single shot target rifles and pistols, especially those using the 22 caliber rim fire ammunition. A 22 caliber rifle or pistol equipped with a silencer is the ideal thing to take along on the fishing, or hunting trip. It will enable you to pick off a squirrel, grouse or rabbit for the larder, or indulge in a little target practice without disturbing others, or frighten the big game you have come many miles to secure. The silencer is especially recommended to those who live in the cities and settled districts who would like to enjoy target practice but are unable to do so on account of others objecting to the noise. By attaching a silencer to your small bore rifle or pistol you will be able to shoot in the basement or indoor range to your own satisfaction without annoying others.

WHAT THE SILENCER DOES

The silencer eliminates the greater part of the noise due to the report of the explosion. It will eliminate more than 75 per cent of the noise when attached to arms using the 22 caliber rim fire ammunition. In the case of the Hi-power rifles the recoil is reduced over two-thirds, as is also the ear splitting report of the explosion. It is the sudden liberation of the powder gas into the open air, after the bullet leaves the muzzle of the gun, that causes the report. This gas, or muzzle blast, is caught in the silencer and caused to expand after which it is slowly released into the air. It is the tendency of this muzzle blast to blow the silencer off the gun that counterbalances the backward thrust, or kick, of the gun. On the automatic or the single shot target pistols it improves the marksmanship to a surprising degree as it eliminates nearly all the recoil or "flip."

A silencer in no way affects the accuracy or the flight of the bullet.

BULLET FLIGHT NOISE

The silencer does not control the noise made by the bullet in its flight through the air beyond the gun. This noise is like the crack of a whip and not a "whistle" or "shriek" as is commonly supposed. If the bullet velocity exceeds 1,075 feet per second the noise will be present. The 22 caliber short and long rim fire ammunition are below this while the 22 long rifle, the 22 W.R.F. and the 22 Hi-Speed and all high power ammunition are above the critical velocity and make a noise known as bullet flight noise. This noise can only reach the shooter's ears by being reflected back from some object, such as the ground, trees or buildings. If you are out in the open air and fire a gun, equipped with a silencer, straight up in the air you will only hear a soft noise, because there is nothing up in the air to reflect the noise back.

HOW TO TELL IF THE SILENCER IS SILENCING

Place the muzzle of the gun, with the silencer attached, within three or four inches of a bag of soft earth or sand and fire the gun. All you should hear would be a heavy "grunt." After trying this step back about twenty-five feet and fire into the bag of sand. A sharp "crack" will be heard. This is the BULLET FLIGHT NOISE. The gun report is the same whether held close to the bag or not. If you doubt it, try it.

THE CONSTRUCTION OF THE SILENCER

The bodies of these silencers are made of seamless steel tubing. Instead of a series of silencing chambers, or cup discs, as is the case with other makes of silencers, the Brown silencer has a series of baffles set at such an angle as to deflect the gases toward the outer walls of the silencer, and from one baffle to the other for the full length of the silencer, finally releasing the gases into the open air, greatly reducing the muzzle blast. This arrangement also permits the greater part of the sediment to be carried away with the muzzle blast. It also permits easy cleaning without the necessity of having to take the silencer apart, it only being necessary to place the silencer in boiling water for a few minutes, remove, shake out the water and stand on end to dry.

By the use of the new Non-Corrosive ammunition very little trouble will be experienced with the silencers filling up with sediment, as was the case with the old black powder ammunition. In the case of the 22 Long rifle Non-Corrosive ammunition many hundreds of rounds have been fired through one of these silencers without any signs of filling up. The 22 long is the worst, being noisy; inaccurate and dirty.

ARMS ON WHICH SILENCERS CAN NOT BE USED

SILENCERS CANNOT BE USED ON REVOLVERS, AUTOMATIC MILITARY OR POCKET PISTOLS. Silencers cannot be used on revolvers on account of the gas escaping through the gap between the cylinder and the barrel, making it impossible to silence them. Automatic pistols and rifles in which the barrels recoil to the rear in order to unlock the breech, such as the Luger, Mauser and the Colt military models and Remington model 8 Hi-power rifles will not work with silencers attached. The forward pull of the silencers renders the action inoperative. Automatic rifles and pistols in which only the breech block travels to the rear, work excellently with a silencer attached.

HOW SILENCERS ARE ATTACHED

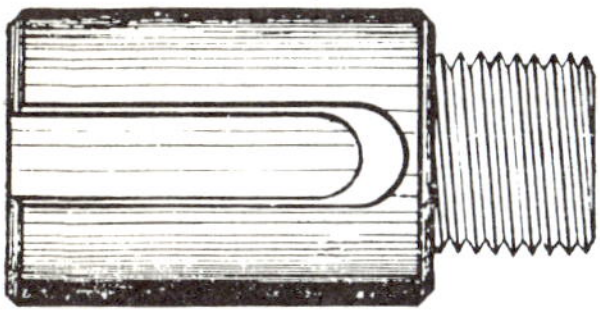

SPECIAL COUPLING

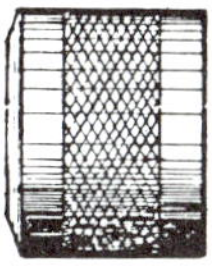

THREAD PROTECTOR CAP

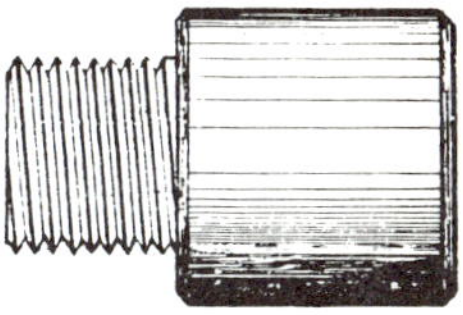

STANDARD COUPLING

The small size silencers for rifles and pistols using the 22 caliber rim fire ammunition are attached to the barrel by a friction coupling driven over the muzzle of the barrel. The silencer being screwed on to the coupling. These couplings are carried in stock for most all standard makes of rifles and pistols made in this country. By the use of extra couplings the same silencer may be used on any number of arms of the same caliber. Such arms as the Colt Woodsman, Hartford and Reising pistols, and the Model 52 Winchester and the M1 Springfield 22 caliber rifles require special couplings on account of the position of the front sight. The coupling has a slot milled in the body to accommodate the sight as shown in the cut at the upper left. The cut at the upper right is that of a plain, or standard coupling. The upper center is a knurled cap to protect the threads of the coupling or barrel when the silencer is removed.

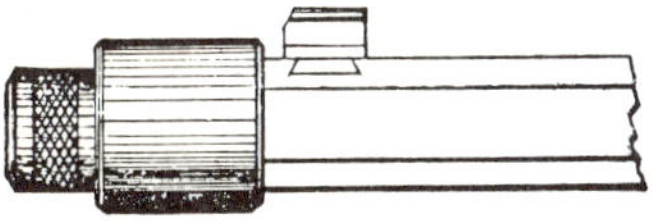

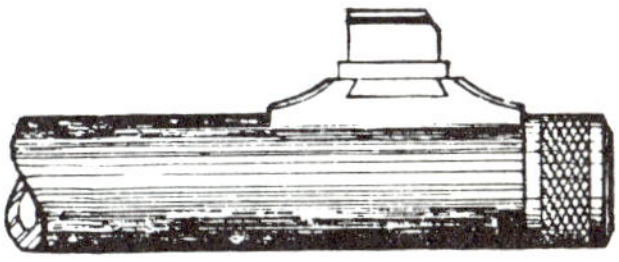

At the left of this paragraph is shown a standard coupling, with thread protector cap, fitted to the muzzle of an octagon barrel. At the right is shown a threaded barrel with thread protector cap in place. Only in the case of some 25-20 and 32-20 rifles does the silencer interfere with the sights. A sheet of instructions for attaching and cleaning the silencer is sent out with every silencer.

WARRANTY

These silencers are all inspected and tested for accuracy before leaving the factory, and are guaranteed against any defects in workmanship or material. Should they prove defective, in any way, within ten days after receipt of same return to factory, postage prepaid, and they will be repaired or replaced, as the case may warrant, free of charge.

The body of the 22 caliber silencer is ⅞ in. diameter ¾ inches long and weighs 5 ozs. The price of $7.00 includes one body with standard coupling and thread protector cap.

Silencer Body for Cal. 22 .... $5.50

Standard Couplings 1.00

Knurled Thread Protector Cap .... .50

Couplings for Colt, Hartford and Reising Pistols 1.50

Couplings for Winchester 52 and Springfield M. I. 2.00

The body of the silencer for arms using the 22 W.R.F., 22 Hi-Speed, 25 rim fire, 25-20 and 32-20 center fire ammunition is 1⅛ in. diameter, 8½ inches long and weighs 11 ozs. The threaded plug is welded into the body and the baffles are welded into a solid core in this and the Hi-power silencer making them almost indestructible.

Price of Silencer with Thread Protector Cap $10.00

The Hi-power silencer is bored for all calibers from the 22 Hi-power to 38 inclusive. It is 1¼ in. diameter and about 10 inches long and weighs 15 oz.

Price of Silencer with Thread Protector Cap $12.00

Barrels on all arms using ammunition larger than 22 caliber rim fire must be threaded in order to attach a silencer. It is dangerous to attach a silencer to a Hi-power rifle with a coupling as they are apt to blow off and be the cause of a serious accident.

Price for Threading Barrels $3.00

NOTE

Be sure to give the make and model of the rifle or pistol when making out order. When sending the rifle in to have the barrel threaded it is better to send in the whole rifle in order that they may be tested out to make sure that the silencer is in perfect alignment with the bore of the barrel. All transportation charges are paid by purchaser. When personal checks are sent in payment of goods the order will be held until the check has had sufficient time to go through the clearing house.

REMARKS

PARKER'S SILENCERS OR SOUND MODERATORS FOR FIREARMS.

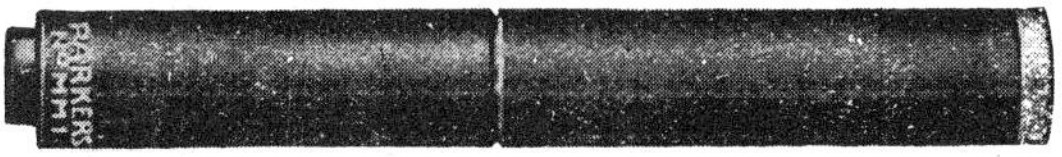

List No. M.M. 1 for Low Power Rifles up to ·25 in Calibre.

Length $6\frac{7}{8}$" Diameter $\frac{7}{8}$" Weight $5\frac{1}{2}$ ozs.

Price with Thread Cap and Assembling Tool .. **25/-**
Our charge for screw cutting barrel **10/-**

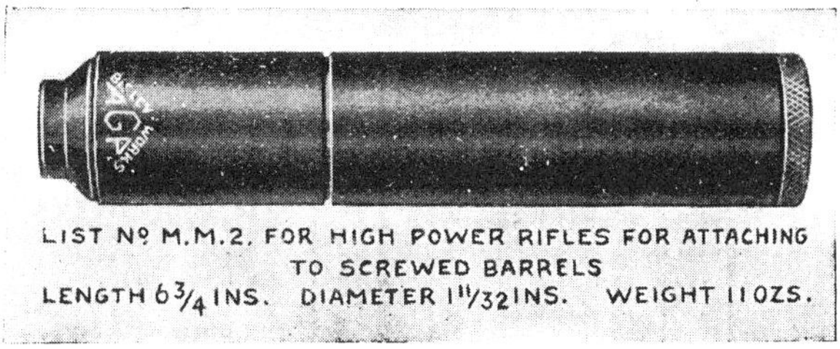

List No. MM.2 for all High Power Single Barrel Rifles up to ·45-in. Calibre.
This model must be attached by screw cutting the barrel.

Length $6\frac{3}{4}$" Diameter 1-11/32" Weight 10 ozs.

Price with Thread Cap and Assembling Tool .. **35/-**
Our charge for screw cutting barrel **10/-**

How the Silencer Operates.

The Silencer checks the muzzle blast. Instead of the powder gases being liberated into the air instantaneously when the bullet emerges from the muzzle, they are caught by the Silencer, and allowed to escape gradually.

This is accomplished by the SILENCING CHAMBERS, which are so constructed as to offer a very high resistance to a rapid flow of gas through them, but very little resistance to a slow flow of gas.

The hole in the Silencer is much larger than the bullet and the latter does not touch anything in passing through, and consequently accuracy of flight and velocity are just the same with or without the Silencer.

How much it reduces the Noise.

The Silencer annuls almost all the noise of the gun report. On high power rifles it also reduces the recoil over two-thirds. The reason for this is that there is a very strong tendency for the Silencer to be blown off the end of the gun. This tendency amounts to a forward pull on the gun barrel. This forward pull counter-balances part of the backward kick.

The only noise the Silencer does not control is the noise made out in the air beyond the gun by a bullet in its flight. This noise is a " crack " like the crack of a whip and not a " whistle " or a " shriek " as is commonly supposed. The noise is made by the same thing that causes the air to crack when a whiplash is snapped.

This noise cannot be avoided when the bullet velocity exceeds 1,075 feet per second, no matter how quiet we make the gun. ·22 Short and ·22 Long Rifle ammunition is below this, and flies quietly. ·22 Long and ·22 W.R.F. and all high power ammunition are above the critical velocity and usually give bullet flight noise.

Shot cartridges up to ·45 calibre work perfectly in the Silencer and make noiseless miniature trap shooting possible without disturbance or danger. On single shot target pistols it improves marksmanship to a surprising degree as it avoids nearly all recoil, or " flip."

The Silencer will not operate on revolvers. No revolver can ever be silenced. If you stop the noise at the muzzle, it makes its escape at the gap between the cylinder and the the barrel. Every revolver must have this gap.

The Silencer is a practical device for use on all Repeating, Automatic, Single Shot Rifles, and Single Shot Target Pistols. It is made in all calibres from ·22 to ·45.

THE PARKER-HALE NEW DETACHABLE COUPLINGS

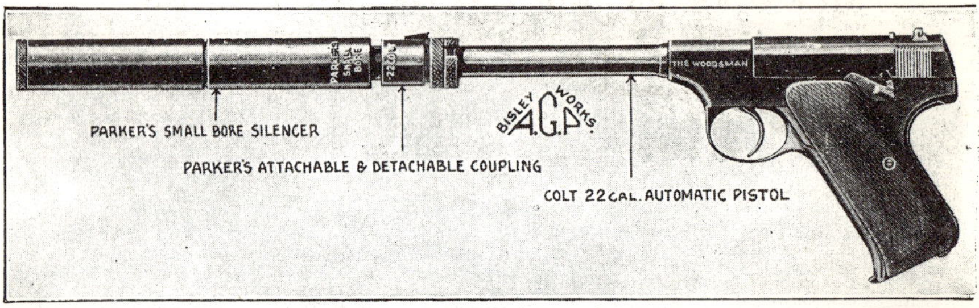

Colt .22 cal. Automatic Pistol fitted with Silencer and Parker-Hale New Detachable Coupling.

The types of coupling shewn on this page have been evolved after a great deal of experience in the fixing of sound moderators and wherever it is practical to use them, they will be found thoroughly efficient, though naturally not so neat as a screw-cut barrel. Couplings however, do help the man who cannot conveniently manage to send us his rifle or to get the necessary lathe work done locally, and these shewn on this page are made to slip on the barrel and fit with the minimum of trouble. It must be realised of course that though they are generally of a standard size, barrels vary in diameter, some to a great extent and others scarcely noticeably; it always pays therefore to give fullest particulars when ordering couplings and in every case to state at least the diameter immediately at the muzzle end, though it is useless to take this measurement other than with a micrometer or other such fine measuring instrument.

Specially designed for the Colt "Woodsman" ·22" cal. automatic pistol, this coupling is fitted by registering the slotted rings with the slot in the Body, slipping the attachment on to the barrel, seating it properly with a mallet and screwing the first ring hard up against the foresight block and locking it with the second (see the complete job above).

List No: MM.10. Price **8/6.**

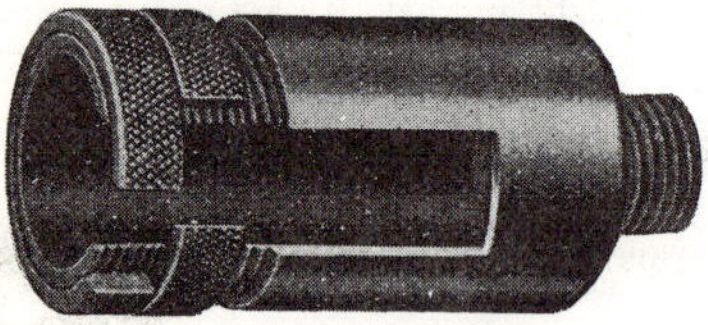

Having a demand from America for couplings to enable the enthusiastic small-bore rifleman to fit a silencer to his target rifle to enable him to practice in his basement without creating domestic trouble, we have adapted the principle of the Colt 'Woodsman" coupling to the popular Winchester "52" rifle. Blanks are also kept in stock for making up to special order at **15/-**

List No: MM.12 Price **12/6**

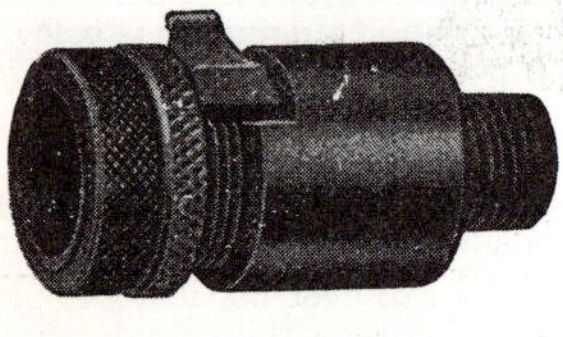

The foresight shewn is not supplied with this coupling; it is superimposed merely to explain the method of fitting this type of coupling to those rifles which are not provided with a convenient foresight block of dimensions suitable for an attachment similar to the two shewn above. When ordering this type a very exact sketch giving measurements in detail is required. It is not stocked in finished condition for any rifle in particular but blanks are kept ready for making up to order.

List No. MM.13. Price **12/6**

PARKER-HALE SOUND MODERATORS ARE FULLY DISCUSSED IN THE LEAFLET PROVIDED FREE ON REQUEST

PARKER'S MODEL 1927
SILENCER for ·410 GAUGE SHOT GUNS.

Having for some years experienced a demand for ·410 gauge Silencers, during which time we were expecting delivery from the original makers, we at length decided to manufacture one in our own works, which we have no hesitation in saying is of a better design than its predecessor. As will be seen in the lower illustration, a long sleeve is fitted to the barrel, either by screwing or soldering, on which the Silencer is detachably fixed.

With 2″ case cartridges the report is reduced by 50% ; with 2½″ cases it is naturally not so silent, but the sharp crack is reduced to a sort of thud which would scarcely be noticed at 200 yards distance, and consequently the disturbance of game would be very much less than is usual.

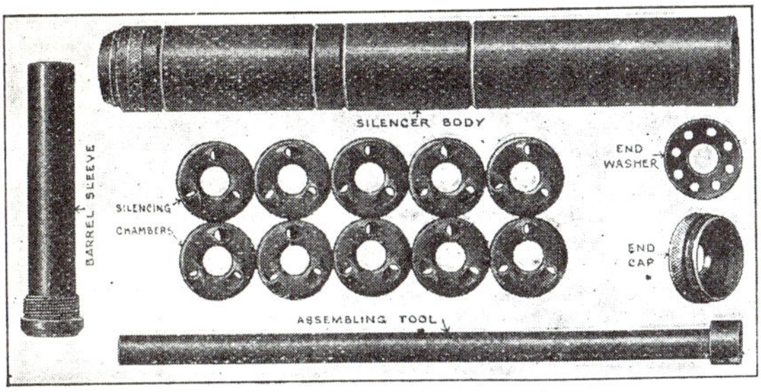

PRICES :

Model 1927 Silencer

As illustrated ..	40/-
Fitting to Gun ..	10/-

N.B.—If customer desires to do the fitting himself, please give the diameter of the barrel at the muzzle and 4″ further back, and we will do our best to send a close fitting sleeve for soldering on.

Testimonial from W. M. Pillans, Esq., Holmwood, Holme, Nr. Peterborough.
9/9/29. "I am very pleased with the ·410 Silencer. The report from the Gun with it on is no more than a ·22 cal. Rifle. Game are not disturbed by the noise."

BRITISH TESTIMONY.

From H. C. M. Shaw.
"The ·22 cal. Silencer works very well. A partridge sat and allowed me to have 4 shots at it, evidently hearing nothing. The last shot killed it."

Received from B. C. Melton, 267, High Holborn, W.C.1. 8-9-26.
"I have received Silencer for which I thank you very much. After trying it over the week end, I find it very effective."

Major F. S. Gillespie, R.A.M.C., British Military Hospital, Kamptee, C.P. India. Received 14-2-27.
"I have had great value out of the Silencer for my ·22 Browning which I got from you, it is invaluable for shooting for the pot when out in the jungle, when the use of a shot gun or ordinary ·22 would disturb game."

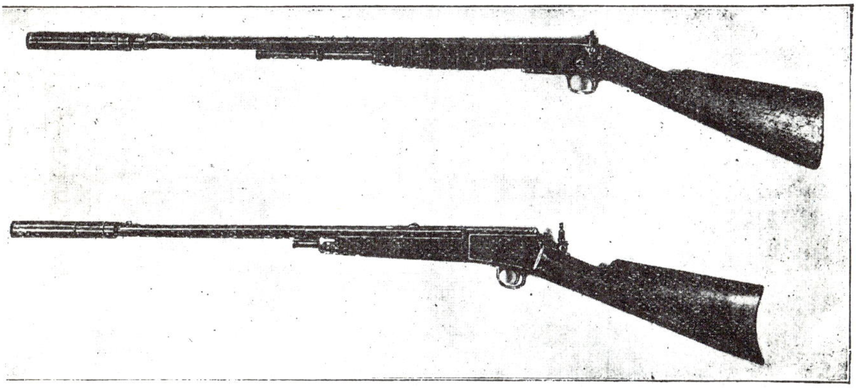

Top illustration shows Silencer fitted on a Remington No. 12 Repeater by means of a coupling. Bottom illustration shows Silencer fitted on a Winchester ·22 Automatic, the barrel being screwed.

HOW THE MODERATOR OPERATES.

The Moderator checks the muzzle blast. Instead of the powder gases being liberated into the air instantaneously when the bullet emerges from the muzzle, they are caught by the Moderator and allowed to escape gradually.

This is accomplished by the baffle plates which are so constructed as to offer a very high resistance to a rapid flow of gas through them, but very little resistance to a slow flow of gas.

The hole in the baffle plates is larger than the bullet and, therefore, the latter does not touch anything in passing through, consequently accuracy of flight and velocity are just the same with as without the attachment.

HOW MUCH IT REDUCES THE NOISE.

It annuls almost all the noise of the gun report. On high power rifles it also reduces the recoil over two-thirds. The reason for this is that there is a very strong tendency for the Moderator to be blown off the end of the gun. This tendency amounts to a forward pull on the barrel. This forward pull counter-balances part of the backward kick.

The noise that the Moderator does not control is the noise made out in the air beyond the gun by a bullet in its flight. This noise is a " crack " like the crack of a whip and not a " whistle " or a " shriek " as is commonly supposed. The noise is made by the same thing that causes the air to crack when a whiplash is snapped. **This noise cannot be avoided when the bullet velocity exceeds 1,075 feet per second, the speed of sound.**

Sound Moderators as a consequence are most efficient when used on small bore rifles such as .22 calibre rim fire, with **ordinary** velocity Short or Long Rifle cartridges, for these bullets fly at a speed less than sound and therefore do not make " bullet flight noise." All the high speed brands of .22 calibre rim fire cartridges make this noise and their use should, therefore, be avoided if silence is desired.

Sound Moderator. List No. MM.1 includes Thread Cap and Assembling Tool.
For low power rifles up to .25″ calibre.
Length 6⅞″ Diameter ⅞″ Weight 5½ ozs.

Persons in the U.K. who require Sound Moderators fitted to their rifles must apply to the police to have their Firearms Certificates varied to enable them to acquire one under the Firearms Consolidation Act of 1937.